MANAGING IoT AND MOBILE TECHNOLOGIES WITH INNOVATIO░ AND S░ ░OMPUTING

T0074589

MANAGING IoT AND MOBILE TECHNOLOGIES WITH INNOVATION, TRUST, AND SUSTAINABLE COMPUTING

Edited by
Kris MY Law
Andrew WH Ip
Brij B Gupta

and

Associate Editor
Shuang Geng

CRC Press
Taylor & Francis Group
Boca Raton London New York

CRC Press is an imprint of the
Taylor & Francis Group, an **informa** business

First edition published 2021
by CRC Press
6000 Broken Sound Parkway NW, Suite 300, Boca Raton, FL 33487-2742

and by CRC Press
2 Park Square, Milton Park, Abingdon, Oxon, OX14 4RN

© 2021 Taylor & Francis Group, LLC

CRC Press is an imprint of Taylor & Francis Group, LLC

The right of Kris MY Law, Andrew WH Ip, Brij B Gupta, and Shuang Geng to be identified as the authors of the editorial material, and of the authors for their individual chapters, has been asserted in accordance with sections 77 and 78 of the Copyright, Designs and Patents Act 1988.

ISBN: 978-0-367-42216-5 (hbk)
ISBN: 978-0-367-75586-7 (pbk)
ISBN: 978-0-367-82275-0 (ebk)

Typeset in Caslon
by SPi Global, India

Contents

Editors

Kris MY Law is an Associate Professor at the School of Engineering, Deakin University, Australia. Prior to her joining Deakin University, she was a Lecturer at the Department of Industrial and Systems Engineering, Hong Kong Polytechnic University. She also holds a Docentship (Adjunct Professorship) in the Department of Industrial Engineering and Management, Oulu University in Finland. She undertook a postdoctoral research scholarship and was a Visiting Researcher at the Graduate Institute of Industrial Engineering, National Taiwan University from 2009 to 2011.

Her expertise lies in organizational learning and development (OLD), technological innovation and entrepreneurship (TIE), engineering education, and smart industrial initiatives (SII). She has been invited as a visiting scholar to Taiwan, Thailand, and Europe (Finland and Slovenia), and she has been active as a professional OL consultant promoting project-based action learning (PAL) in high-tech organizations. Her publications include books, book chapters, and journal articles (SCI and SSCI indexed).

Andrew WH Ip has more than 30 years of experience in teaching, research, education, industry, and consulting. He earned a PhD at Loughborough University (UK), an MBA at Brunel University (UK), and an MSc in industrial engineering at Cranfield University (UK) and received an LLB (Hons) from the University of Wolverhampton (UK). He is Professor Emeritus of Mechanical Engineering at the University of Saskatchewan and a Principal Research Fellow in the Department of Industrial and Systems Engineering at the Hong Kong Polytechnic University. He is also a visiting professor at various universities in mainland China and an Honorary Industrial Fellow at the University of Warwick, Warwick Manufacturing Group, UK.

Professor Ip has published nearly 250 papers, with over 150 papers in SCI-indexed journals, and more than 130 papers at conference proceedings as well as books and book chapters. He is also the Editor-in-Chief of *Enterprise Information Systems*, Taylor & Francis (SCI Indexed), and the Editor-in-Chief and Founder of *International Journal of Engineering Business Management*, SAGE (ESCI- and SCOPUS-indexed), and an editorial member of various international journals. He is a senior member of the Institute of Electrical and Electronics Engineers (IEEE) and a member of the Hong Kong Institution of Engineers (HKIE).

Brij B Gupta earned a PhD at the Indian Institute of Technology, Roorkee, India, in the area of information and cyber security. In 2009, he was selected for a Canadian Commonwealth Scholarship award by the Government of Canada ($10,000). He has published more than 250 research papers (including 8 books and 22 book chapters) in international journals and conferences of high repute, including IEEE, Elsevier, ACM, Springer, Wiley, Taylor & Francis, Inderscience, etc.

He has visited Canada, Japan, Australia, China, Spain, Hong Kong, Italy, Malaysia, Macau, etc. to present his research. Dr. Gupta is a Senior Member of IEEE, Member ACM, SIGCOMM, SDIWC, Internet Society, Institute of Nanotechnology, Life Member, International Association of Engineers (IAENG), Life Member, and International Association of Computer Science and Information Technology (IACSIT).

Shuang Geng is an Assistant Professor at the College of Management, Shenzhen University, China. She earned a PhD at the Department of Systems Engineering and Engineering Management, City University of Hong Kong, Hong Kong, SAR. She was a visiting scholar at the College of Management, Bath University, UK.

Her research interest includes intelligent recommendation system, optimization, data analytics, and technology enhanced learning. Her papers appear in *Computers and Education, Internet Research, British Journal of Education Technology, Project Management Journal, Industrial Management and Data Systems*, etc. She was the Session Chair of the 16th International Conference on Service Systems and Service Management (ICSSSM'19).

Contributors

YC Chau
System Engineering and
    Engineering Management
    Department
City University of Hong Kong
Kowloon, Hong Kong

Pooja Chaudhary
National Institute of Technology
Kurukshetra, India

Vincent WC Fung
Industrial and System
    Engineering Department
Hong Kong Polytechnic
    University
Hong Kong SAR, China

Shuang Geng
College of Management
Shenzhen University
Shenzhen, PRC

Brij B Gupta
Department of Computer
    Engineering
National Institute of Technology
Kurukshetra, India

Miaojia Huang
College of Management
Shenzhen University
Shenzhen, PRC

Kris MY Law
School of Engineering (SEBE)
Deakin University
Victoria, Australia

Dragan Peraković
Department of Information and
    Communication Traffic
University of Zagreb
Zagreb, Croatia

Konstantinos Psannis
School of Information Sciences
University of Macedonia
Thessaloniki, Greece

Chui Kwok Tai
Open University of Hong Kong
Kowloon, Hong Kong

Yingsi Tan
College of Management
Shenzhen University
Shenzhen, PRC

Chenyu Xu
College of Computer Science
   and Software Engineering
Shenzhen University
Shenzhen, PRC

Lu Yang
HKU Business School
The University of Hong Kong
Hong Kong, China

1

Communication between Human and Machines in the Era of Industry 4.0

BRIJ B GUPTA, POOJA CHAUDHARY,
KWOK TAI CHUI,
AND KONSTANTINOS PSANNIS

Contents

1.1 Industry 4.0 (I4): Essence in the Nutshell

Evolution of technology has led to significant improvement in human–machine interface (HMI) [1]. From simple push button to touch-screen display, the industry has witnessed rapid growth in technology. Still, many advancements of HMI are yet to come. With the evolution of Industry 4.0 (I4), organizations have witnessed fast, more flexible, efficient, and autonomous productions. It was first termed by the BMBF in Germany in the year 2011, in which they have demonstrated how cyber–physical systems (CPS) could bring advancements in business models which could bring a paradigm shift in the industrial automation sector [2].

1

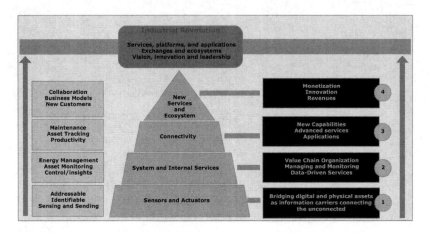

Figure 1.1   Industrial transformation pyramid with Industry 4.0.

I4 represents the network of self-regulating and automated machinery that possesses the capabilities to interact with other devices/processes, enabling advanced path of production. It revolutionizes the data-intensive manufacturing procedures into more connected ecosystem of smart devices to realize the vision of smart industry. Specifically, it reconciles the physical devices/machinery in industries and digital technologies into CPS. Figure 1.1 depicts how I4 assists in industrial transformation.

1.1.1 Evolution of Industry 4.0

I4 refers to the complete digitalization of manufacturing industries and requires no or less human intervention. Moreover, various technologies form the basis of this growing trend such as Internet of Things (IoT), cloud computing, Big Data, cognitive computing, and artificial intelligence (AI) [3,5–9]. Nevertheless, before going into detail of industrial 4.0, let's dive into its progression since Industry 1.0. The four stages of industrial revolution are depicted in Figure 1.2.

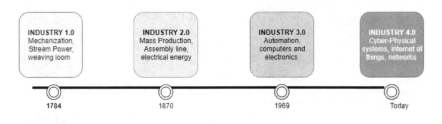

Figure 1.2   Four stages of industrial revolution.

- **Industry 1.0:** Mechanization was at the root of Industry 1.0 which contributed to modern production as the way we know it. The invention of the first mechanical loom in 1784 had marked the beginning of the first phase of Industry 1.0 and the industrial revolution. Later, it is followed by the invention of water and steam-driven machines that allow staff to run machinery day and night without numerous operators being required.
- **Industry 2.0:** Mass production of goods by assembly lines and labour division were characterized in the era of Industry 2.0 to optimize the workforce. Electricity has appeared at the beginning of the 20th century which replaced water and steam, thus allowing companies to concentrate their power sources on their individual machines.
- **Industry 3.0:** In the early 1970s, the third phase of the Industrial Revolution began with the development of the programmable logic controller (PLC). Industry 3.0 lies behind the broad use of electronics and information technology for the advancement of digitalized manufacturing and automation.
- **Industry 4.0:** At present, we are witnessing the fourth Industrial Revolution, next phase in automation and digitalization of manufacturing, which promises to maintain the productivity growth and improvements in industrial functions. It is estimated that through adopting I4, organizations could increase their revenues by $493 billion annually and cut costs by $421 billion annually by 2020.

In accordance with I4, integrated processing and communication capabilities are used for all objects in the world of factories. This not only affects machine-to-machine (M2M) communication but has wide-ranging consequences for human and technological interactions. In perspective of technological advancement, the range of problems as well as demands for human power will change in the factory space. The use of automated and self-organizing devices and/or processes will transform the complex manufacturing processes into fast, efficient, and feasible processes that yield high-quality goods at reduced production cost. There are essentially nine technologies that model I4, as highlighted in Figure 1.3.

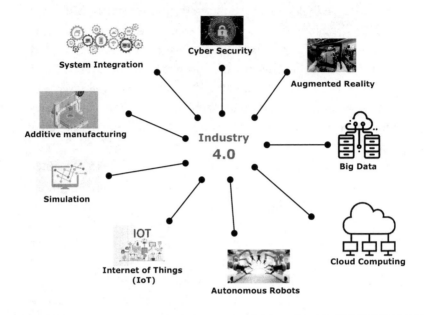

Figure 1.3   Technological components of Industry 4.0.

1.1.2 Cyber–Physical System (CPS): Basic Units of Industry 4.0

The key component or the basic unit of I4 is a CPS. It is a mixture of computational and physical systems. CPSs have better intelligence and the ability to communicate in real time with other similar systems when compared with current machines. The integration of sensors and actuators helps to control the movement of the machine, to sense the variations in its surroundings and to exchange information with other system via the network. In one sentence, CPS connects physical and digital system in industrial perspective [10–13]. It facilitates interaction between industrial systems and makes them connected with each other to enable new capabilities; thereby, it helps in realizing the vision of smart health, smart factory, smart cities, smart logistics, and so on, as shown in Figure 1.4. However, compared to the computer integrated manufacturing (CIM) strategy, I4 does not shift towards production facilities without jobs. Rather, individuals should be incorporated into the cyber–physical structure to fully realize their individual skills and abilities. The human–cyber–physical system (HCPS) relationship is further divided into a virtual digital component and a physical component. The interaction between human and

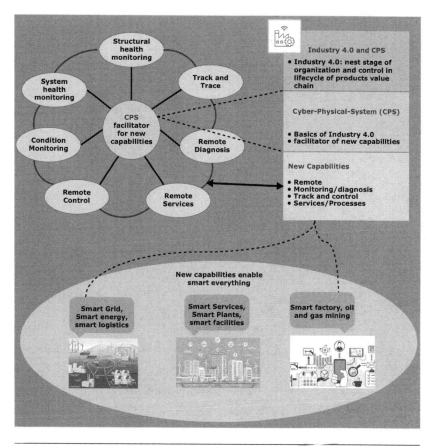

Figure 1.4   CPS: Enabler of new capabilities.

CPS takes place either through direct control or via user interface. Such a close interaction between human and CPS also raises socio-tech questions about self-reliance and decision-making.

The decision-making and control procedures by the human can be carried out on-site or from a wide variety of different manufacturing complexes. It assumes that more responsibilities and growing activities are expected by the individual worker in future. In fact, when addressing complex problems, the worker is the last instance within the cyber–physical structure to play the role of the innovative problem solver, as depicted in Figure 1.5 [14]. Therefore, to allow human–machine interaction HMI is introduced. Its development is motivated by the need for reliable hardware, improved efficiency, mobility, data security, remote resources, and compliance. The demand for high

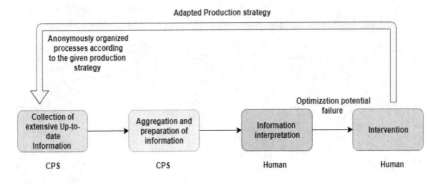

Figure 1.5 Human as a monitor for production strategy and decision-making.

production compels companies to implement agile, smarter, and innovative ways of fostering output through technologies that supplement and increase human labour with robots and decreases industrial accidents that are caused by a process failure.

To gather and analyse data, HMI employs sensors, robots, software, wireless systems, and M2M learning techniques. The collected data are used to perform managerial operations that connect production facilities, corporate back-office and on-site robots, as well as suppliers across the supply chain. However, the focus of integrated intelligence and automation is to limit the interaction of human with the operations and hence to enhance the quality of interaction so that the things respond more efficiently. This leads to the rapid growth in the implementation of AI within the production lines or manufacturing systems, obsoleting the HMI technology, yet the importance of HMI increases when systems cannot be handled automatically.

1.2 IoT Linking Human and Machines

If we speak of the term 'innovation' and where the technology currently takes place then we cannot look beyond the IoT. IoT refers to advanced connection of devices, systems, and services that push/go beyond the boundaries of M2M communication. Simply, IoT is an infrastructure which is well-connected and where virtually every machine or device has basic, integrated intelligence used for data transmission and interaction with other machines/devices and with humans. Figure 1.6 shows the additional layers of IoT ecosystem [10,15–17].

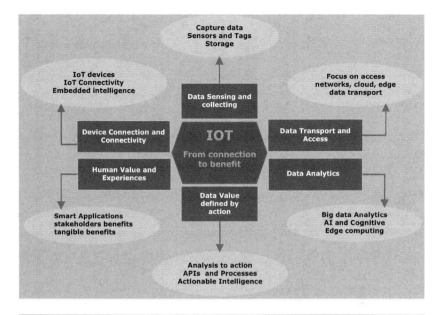

Figure 1.6   IoT: from connection to benefit.

IoT combines devices, sensors, actuators, and humans to enable a free flow of conversation between devices and humans. The advancement of AI and machine learning, the conversation among devices/machines, allows the machine/devices to react, respond, anticipate, and advance the physical world similar to information enhancement which was done by the Internet using the computer screens and network. Most of the current development in IoT is towards the industry [18]. Sensors play an important role in the IoT world. These are the components that are connected to devices to sense the physical world. It measures the physical inputs and converts them into raw data, which is stored for analysis and access. The advancement of sensors helps the devices to understand consumers at every level; hence, the term smart object is coined up. Examples include temperature and thermal sensor, light and imaging sensors, velocity and motion sensors, proximity sensor, pressure and force sensor, and many more.

There is a rapid increase in data because of IoT which is accessible to create informed decisions. Whenever human decision is needed, AI combined with HMI plays a critical role within I4 for time-sensitive and critical decision-making processes. At its core, IoT has removed the friction from everyday tasks and manual operations and thus enabled people to spend time on the things they enjoy by limiting humans to take strategic decisions rather than involving man power.

1.2.1 IoT in Industry: Industrial Internet of Things (IIoT)

Another significant segment of the IoT is the Industrial Internet of Things (IIoT). It is the blend of information technology and operational technology that finds its utility not even in heavy industries such as manufacturing, oil and gas, and transportation, but also in less heavy such as smart irrigation, smart city, smart health, and so on. It is defined by the Industrial Internet Consortium (IIC) as 'systems, devices and people that facilitate intelligent industrial operations, bringing new transformation in industries using advance computing, low cost sensing, data analytics and connectivity'. Figure1.7 represents the key elements of IIoT.

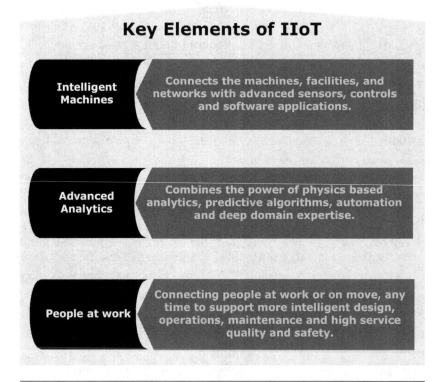

Key Elements of IIoT

Intelligent Machines — Connects the machines, facilities, and networks with advanced sensors, controls and software applications.

Advanced Analytics — Combines the power of physics based analytics, predictive algorithms, automation and deep domain expertise.

People at work — Connecting people at work or on move, any time to support more intelligent design, operations, maintenance and high service quality and safety.

Figure 1.7   Key elements of IIoT.

There are many organizations that are applying IIoT in its operations and many more are in queue [19,20]. It yields multiple benefits to the organizations such as:

- Improve operational efficiency.
- Improve productivity.
- Creates new business opportunity.
- Maximize asset utilization.
- Reduce downtime.
- Reduce asset lifecycle costs.
- Enhance worker safety.
- Enhance product innovation process.
- Provide better understanding of customer demand.

1.2.2 Consumer Internet of Things (CIoT)

There are many applications where IoT has placed its roots and even flourishing every day. One such application is found in consumer electronics and the term Consumer Internet of Things (CIoT) is coined up. Most popular example is smart home. Its applications may range from simple devices like smart watches to complex devices like smart home automation system. It is dissimilar to IIoT only in terms of devices and applications, the technology that enables them and their motive, but at the core it is only IoT that plays its role and there exists many similarities like differences. There are three features that the user expects from their smart devices: usefulness, convenient, and innovative. There are numerous applications of smart consumer electronics, but top 10 applications are as follows:

- Smart home
- Wearables
- Smart city
- Smart grid
- Industrial Internet
- Connected cars
- Connected health
- Smart retail
- Smart supply chain
- Smart farming

1.3 Different Perspectives of IoT – Human, Operation, Managerial

The key aspect of the IoT is to unite physical domain, digital platform, and humans through network of connected processes and
data which turned into useful information and actions. Many people might have thought of explosive growth of sensors and communication devices – that is, the IoT. Now, the IoT encompasses
nearly every aspect of human interest and life. These IoT devices
track and report on our equipment, houses and buildings, our cars
and climate, and many aspects of our cities, worlds, oceans, and
space. These have started to play a vital role in monitoring our
health, for healthy and safer well-being, for our comfort and entertainment, for our financial activities and in many other aspects of
life. In broader prospective, there are many characteristics of IoT
as represented in Figure 1.8.

The pace of development for new sensor types and devices is already
quite fast. Nonetheless, the more critical and challenging role is the
challenge of interoperating and integrating data and information.
To this end, one project, called Semantic Gateway as Service (SGS),

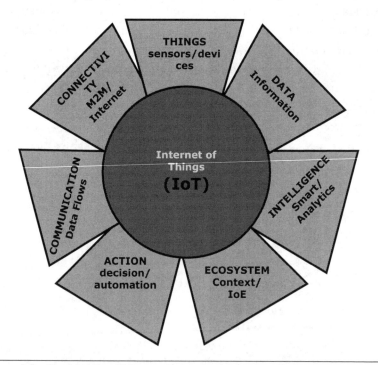

Figure 1.8   Characteristics of IoT.

facilitates the translation of a range of current-use IoT messaging protocols, such as the Extensible Messaging and Presence Protocol (XMPP), the CoAP, and MQTT. Another essential interoperability functionality is provided by the Semantic Sensor Network (SSN) ontology and annotation system of the World Wide Web Consortium (W3C). It is useful to identify any sensor or computer and its data in a standard form, and to help semantic sensor data annotations, rendering the data more meaningful. This essentially provides semantic interoperability between carrying messages.

1.4 Use of Mobile Technologies in IOT

Nowadays, everyone is holding a mobile device in their hand which consists of many features like geo-location and multiple connectivity options like WIFI, Bluetooth, Near Field Communication (NFC), and 5G technologies. In this generation, the technology changed the way humans interact with machines and even in the M2M interaction. Now, we are using mobile apps instead of desktop websites. To simplify the human efforts and interaction with machines, mobile apps are integrated with IOT which can be described as a network of Internet-enabled devices in which every device can be identifiable by a unique IP address and can communicate with each other through the instructions given by the user from the mobile app by smartphone interface. Smart Home is the best example of the integration of mobile application. Some of the basic and major systems of the house such as lighting, air conditioning, security, and even cooking devices are intelligent, meaning that they are able to receive instructions from the user and act accordingly. How all the devices are connected and how does this work? All the device applications that are involved in the network system are interconnected through cloud-based technology to a main hub that can be controlled through an app. This app will be available on the user's smartphone or tablets.

1.4.1 5G Technology

The next-generation mobile standard is known as 5G, the 5th generation technology which is known to create a new platform which is faster and reliable than 4G which is known as Long-Term Evolution (LTE).

Every new generation promises more speed than the previous generation of technology and also loaded with many more advanced features as follows:1G bought us to use the very first cell phones; 2G lead us to text for the first time; 3G bought us to be online; and 4G delivered the speed of data transfer that we enjoy today. As we know that data transfer rate becomes more important when it comes to send or receive any data. Nowadays, the users who use the Internet are increasing rapidly and at the same time the demand of the bandwidth is also increasing. So, to acquire the required bandwidth to the user, here comes the 5G technology which focusesupon three major areas listed below:

- **Massive IoT:** This name is given to the process of bringing vast arrays of previously standalone objects online which makes possible of the smart cities.
- **Critical communication:** This refers to the services that are more important and use cases that cannot simply fail, like digital healthcare and emergency services. So, these require both maximal bandwidth and reliability.
- **Enhanced mobile broadband:** This provides the users to access the superfast broadband on their mobiles which allows them to access any kind of digital services as well as communicate with far higher speed and reliability.

NB-IOT and LTE-M belong to the family of licensed spectrum LPWA technologies which form Mobile IOT to provide support to 5G use cases. The 3GPP (3rd Generation Partnership Project) has confirmed that LPWA use cases will continue which are addressed by NB-IOT and LTE-M which comes under the part of 5G standards. Now, the question arises that how this 5G Network is possible? So, to realize 5G network, five brand new technologies emerged as a foundation: Millimetre waves, Small Cell Networks, Massive MIMO, Beamforming, and full Duplex.

- **Millimetre waves:** Every smartphone and the electronic devices in our home use a frequency under 6 GHz on the radio-frequency spectrum but as the users and the devices getting increased the spectrum is becoming incompatible to serve all the users at a required speed. The solution is to use the licenced band which are over 6 GHz, i.e. around 30–300 GHz. This spectrum has never being used for mobile devices before. But the problem is millimetre waves cannot travel through walls and other obstacles, which lead to the need of the small cell networks.

- **Small Cell Networks:** As the millimetre waves cannot travel through obstacles the base stations should be nearer to each other. When a user is moving from one place to the other, this base station will be automatically shift to the other which has a higher strength of the signal.
- **Massive MIMO:** MIMO means Multiple–Input–Multiple–Output. In today's 4G base stations, we have a greater number of antennas that handle cellular traffic, but when we consider massive MIMO base stations can have about 100 ports which will increase the capacity of today's network to a factor of 22 or more. But, massive MIMO comes with its own complications. The cellular antennas that we use today broadcast information in every direction at once and all those signals will cause crossing of signals and serious inference which lead us to bring up new technology called Beamforming.
- **Beamforming:** It is like a traffic signalling system for cellular signals. So, instead of broadcasting, it allows a base station to send a required stream of data to a specified user. Massive MIMO base station will keep track of the timing and the direction of the signals arriving which makes possible to find the source of the signal.
- **Full Duplex:** Full duplex is the channel, in which sender and receiver can both transmit the data at the same time.

1.5 Future Prospects of Industry 4.0

I4 is the result of automation and digitally data interchange in the manufacturing methods and techniques including IoT, cloud computing, AI, CPS, and various technologies as its backbone. It is an evolving field that demands improvement in every stage of its realization; therefore, there is need to address the challenges that exist in the path of the improvement and can be described as the future prospects of I4. These are pointed out below:

- *Manage complicated system and processes*: to coordinate and regulate the complex manufacturing process, there is need to initiate proper planning for upcoming system and develop appropriate models for existing ones.

- *High-quality network infrastructure*: to handle massive data exchange among smart devices in the industries, we require a reliable and efficient communication network with increased broadband services.
- *Security and privacy concerns*: as I4 is the blending of numerous novel technologies like IoT, therefore, it is vulnerable to the security and privacy issues that are present in these technologies. Adversary might compromise devices that cause severe consequences.
- *Standards and regulations*: as I4 is an evolving field, hence there is a lack of common standards and regulation to assist in collaboration and reference module that provides the information of these standards and regulation.
- *Lack of skill and sufficient qualification*: existing skill sets in the industries are not efficient to incorporate all the technologies required for I4 in the manufacturing process; therefore, it requires proper training of the employees for its working. It is the basic requirement to expedite transformation towards I4.

1.6 Summary

Innovative technologies and digitalization of every field cause the emergence of new industrial revolution, i.e. I4. It embraces multiple novel technologies comprising IoT, AI, cloud computing, Big Data analytics, and so on. It reconciles the physical devices/machinery in industries and digital technologies into CPS. Therefore, this chapter provides a brief overview of I4, its evolution from Industry 1.0, and future research prospects in the field of I4. The main focus of this chapter is to elaborate the key driving technology of I4, i.e. IoT. Furthermore, it provides information on communication ease and development in IoT environment through describing the mobile technologies of IoT.

References

1. Khvoynitskaya, S. (2018). Applications of Advanced Human-machine Interfaces in the Industry 4.0 Era. Retrieved from: https://www.itransition. com/blog/applications-of-advanced-human-machine-interfaces-in-the-industry4-era.

2. Strukturwandel. (2011). Industry 4.0: On the Way to the 4th Industrial Revolution with the Internet of Things. Retrieved from: https://www.ingenieur.de/techik/fachbereiche/produktion/industrie-40-mit-internet-dinge-weg-4-industriellen-revolution.

3. Gupta, B., Agrawal, D. P., & Yamaguchi, S. (Eds.). (2016). *Handbook of Research on Modern Cryptographic Solutions for Computer and Cyber Security.* Hershey, PA: IGI Global.

4. GSMA IoT. (2018). Mobile IoT and 5G: Complementary Technologies central to the IoT. Retrieved from [15]: https://www.gsma.com/iot/news/mobile-iot-and-5g-complementary-technologies-central-to-the-iot.

5. Tewari, A., & Gupta, B. B. (2017). Cryptanalysis of a novel ultra-light-weight mutual authentication protocol for IoT devices using RFID tags. *The Journal of Supercomputing,* 73(3), 1085–1102.

6. Stergiou, C., Psannis, K. E., Gupta, B. B., & Ishibashi, Y. (2018). Security, privacy & efficiency of sustainable cloud computing for big data & IoT. *Sustainable Computing: Informatics and Systems,* 19, 174–184.

7. Gupta, B. B., & Agrawal, D. P. (Eds.). (2019). *Handbook of Research on Cloud Computing and Big Data Applications in IoT.* Hershey, PA: IGI Global.

8. Adat, V., Dahiya, A., & Gupta, B. B. (2018, January). *Economic incentive based solution against distributed denial of service attacks for IoT customers.* In *2018 IEEE International Conference on Consumer Electronics (ICCE)* (pp. 1–5). IEEE.

9. Gupta, B. B., & Gupta, A. (2018). Assessment of honeypots: Issues, challenges and future directions. *International Journal of Cloud Applications and Computing (IJCAC),* 8(1), 21–54.

10. Al-Sharif, Z. A., Al-Saleh, M. I., Alawneh, L. M., Jararweh, Y. I., & Gupta, B. (2020). Live forensics of software attacks on cyber–physical systems. *Future Generation Computer Systems,* 108, 1217–1229.

11. Fiorini, R. A. (2020). Computational intelligence from autonomous system to super-smart society and beyond. *International Journal of Software Science and Computational Intelligence (IJSSCI),* 12(3), 1–13.

12. Hossain, K., Rahman, M., & Roy, S. (2019). Iot data compression and optimization techniques in cloud storage: current prospects and future directions. *International Journal of Cloud Applications and Computing (IJCAC),* 9(2), 43–59.

13. Haji, A., Letaifa, A. B., & Tabbane, S. (2018). New architecture for virtual appliance deployment in the cloud. *International Journal of High Performance Computing and Networking,* 12(4), 362–367.

14. Gorecky, D., Schmitt, M., Loskyll, M., & Zühlke, D. (2014). *Human-machine-interaction in the industry 4.0 era.* In *2014 12th IEEE international conference on industrial informatics (INDIN)* (pp. 289–294). IEEE.

15. Al-Qerem, A., Alauthman, M., Almomani, A., & Gupta, B. B. (2020). IoT transaction processing through cooperative concurrency control on fog–cloud computing environment. *Soft Computing,* 24(8), 5695–5711.

16. Gupta, B. B., & Tewari, A. (2020). *A Beginner's Guide to Internet of Things Security: Attacks, Applications, Authentication, and Fundamentals.* London: CRC Press.

17. Joshi, R. C., & Gupta, B. B. (Eds.). (2019). *Security, Privacy, and Forensics Issues in Big Data.* Hershey, PA: IGI Global.

18. Nendick, J. (2019). How Human–Machine Interaction can Unlock Possibilities in Media and Entertainment. Retrieved from: https://www.ey.com/en_gl/tmt/how-human-machine-interaction-can-unlock-possibilities-in-media.

19. Bharathi, R., & Selvarani, R. (2019). Software reliability assessment of safety critical system using computational intelligence. *International Journal of Software Science and Computational Intelligence (IJSSCI)*, 11(3), 1–25.

20. Sambrekar, K., & Rajpurohit, V. S. (2019). Fast and efficient multiview access control mechanism for cloud based agriculture storage management system. *International Journal of Cloud Applications and Computing (IJCAC)*, 9(1), 33–49.

21. Rossi, B. (2016). How Industry 4.0 is Changing Human-technology Interaction. Retrieved from: https://www.information-age.com/industry-4-0-changing-human-technology-interaction-123463164.

22. Blaisdell, R. (2018). IoT: The Human–Machine Interaction Unlocking New Opportunities. Retrieved from: https://rickscloud.com/iot-the-human-machine-interaction-that-unlocks-new-opportunities.

23. Desai, P., Sheth, A., & Anantharam, P. (2014). Semantic Gateway as a Service Architecture for IoT Interoperability. Retrieved from: http://xxx.tau.ac.il/abs/1410.4977.

24. Compton, M., Barnaghi, P., Bermudez, L., GarcíA-Castro, R., Corcho, O., Cox, S. & Huang, V. (2012). The SSN ontology of the W3C semantic sensor network incubator group. *Journal of Web Semantics*, 17, 25–32.

25. O'Reilly, T. (2014). IoTH: The Internet of Things and Humans. Retrieved from: https://www.oreilly.com/radar/ioth-the-internet-of-things-and-humans.

26. Sheth, A. (2009). Computing for human experience: Semantics-empowered sensors, services, and social computing on the ubiquitous web. *IEEE Internet Computing*, 14(1), 88–91.

27. Tech Ahead Team. (2019). Why Future of Mobile Apps Includes IoT and Bots. Retrieved from: https://www.techaheadcorp.com/blog/future-mobile-apps-includes-iot-bots.

28. Rao, R. (2018). 5G the Next Wave of Wireless. Retrieved from: https://techutzpah.com/5g-the-next-wave-of-wireless.

2

PRIVACY CONCERNS AND TRUST ISSUES

BRIJ B GUPTA, POOJA CHAUDHARY, DRAGAN PERAKOVIĆ, AND KONSTANTINOS PSANNIS

Contents

2.1 Trust Models and Architectures

Internet of Things (IoT) refers to the distributed network of inter-connected heterogeneous physical devices which is a complex environment that cannot be trusted intuitively as these devices can be controlled and monitored by anyone at any-time anywhere around the world [1,2]. This raises the question: 'How can an ordinary user trust the device's security and can check the integrity of the installed software?' In 2017, a survey report has unveiled that the 72% of IoT consumers are unaware about their device being in the compromised state [3–6].

Most prominently, the IoT consumers must be able to interact with such a massive and complex network of IoT devices in a secure manner.

As device manufacturers are more focused towards the infrastructural development and are less engrossed towards the device's safety, therefore, to facilitate robust IoT security, there has to be an efficient solution to answer how people can trust the privacy and security of changes brings out by IoT development. This transformation is heading in right direction if people should be able to entrust IoT devices in the similar manner as they do on purely physical devices. Hence, it appeals for the development of solution that helps in understanding and make normal people aware of device's integrity. Consequently, it leads to the designing of trust models for IoT environment that is efficient as well as broadly accepted [7]. Trust model aids in describing the following aspects:

- How to protect and govern the IoT resources?
- How people can rely on device's ability to maintain security and privacy?
- How can device possess the capabilities to protect themselves from being attacked?

More precisely, trust model illuminates the responsibilities of device vendors and service providers towards providing secure devices so that user can use them without agitation. Here, 'trust' indicates reliance; thus, trust model represents how one entity can rely on another entity in a network. However, if trust model is designed specifically for professionals then it may not be that effective as if it is designed by keeping average users in consideration. This is so because end users are mostly affected by an attack eventually. Hence, trust model should be human centric so that user can take some responsibility of its own security. It will be capable of resolving the following issues:

- If a user gives access to its own device to others, then how can user be able to trust them?
- How can user limit device's usability by others?
- On what a user can trust for protection?

For successful development of secure devices, the system design should be directed by the trust model to monitor the communication and functionality of device. It also supports the effective up-gradation of firmware software of the smart devices. To accomplish the designated functionalities, trust model consists of various components.

The selection of these components must be completed properly and thereby, the following points should be taken into account:

- **Smart devices and embedded software:** it includes various aspects such as features of device provided by the vendor for security and privacy, device trustworthiness, and how devices can communicate with each other. So these factors must be taken into consideration while designing trust model.

- **Resources:** there are various resources or services offered by devices via Internet such as sensing information, status information, controlling information, and so on. How can users know about these resources and how do they control access to device? It is very difficult to make this information intuitive to most of the people.

- **Features reliability:** the data collected by the smart devices are processed for further analysis. How can these data be labelled safely so as not to reveal the identity of the user? How to ensure the correct usage of these labels? Techniques such as classification and data anonymization are used for this purpose. How to ensure their trustworthiness? Thus, these factors should be addressed while designing trust model for IoT.

- **Trust transference:** one aspect which should be addressed by the trust model is trust delegation. It means what are the implications if user provides access to its own device to others. How can this be completed reliably? How trust transference can be achieved by the user? How can user control access to device information by the others?

- **Performance assistance:** it includes methods that aids in understanding the effects of the some actions such as trust transference. It helps in dealing with scaling and complex nature of IoT environment.

- **Authentication system:** for the entire system to be safe, device authentication is must. It is required for correct and secure execution of assigned task. It is a challenging task to develop proper system for identity verification and make this intuitive [8–13].

The generic trust model for IoT environment is depicted in Figure 2.1. This model may be served as the basic model for further enhancements.

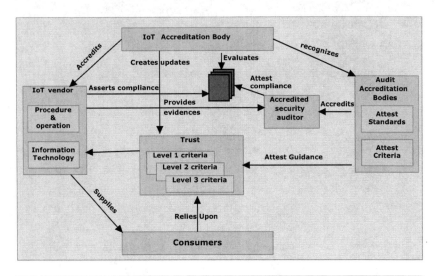

Figure 2.1   Generic trust model for IoT.

2.2 Types of IoT Authentications

With the advent of the IoT, there has been an exponential growth of innumerable cyber-attacks. Some are initiated by the infecting the legitimate IoT devices but most of the attacks are initiated through masqueraded IoT devices. Therefore, strong authentication mechanisms are the demand of an hour, so that devices can be trusted and their identity can be verified [14]. While designing an effective authentication method, researchers should address the threats that may occur in the operational life cycle phases of the smart devices. Due to unique characteristics of the IoT environment, existing authentication methods cannot be applied. Researchers have designed various authentication mechanisms [14,15] specifically for IoT which may be classified into various categories on the basis of their similar attributes as shown in Figure 2.2.

1. **Authentication factor:** it means the factors used for authenticating the IoT device. It is divided into two categories:
   - **Identity:** it represents the information provided by one entity to other entities for verifying its original identity. These authentication methods may include hashing, symmetric, and/or asymmetric cryptographic algorithms.

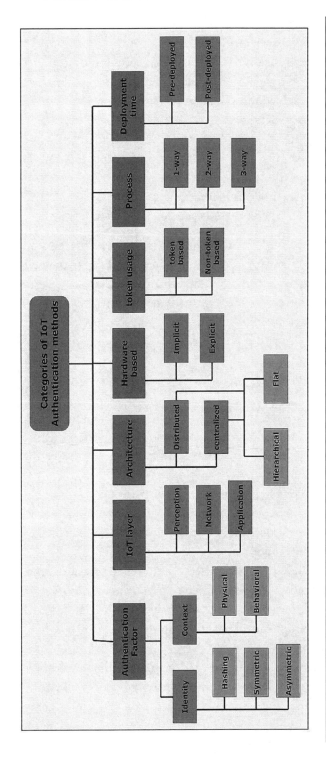

Figure 2.2   Categories of IoT authentication methods.

- **Context:** it indicates the unique information of any entity. It may be:
  a) Physical: it represents the unique biometric features of any person. For example, fingerprints, iris scan, etc.
  b) Behavioural: it represents the behavioural features of any person like typing speed, walking style, etc.
2. **IoT layer:** it represents the IoT layer at which authentication mechanism is working. It may be applied at perception layer, network layer, or application layer.
3. **Authentication architecture:** it describes the designing structure of the authentication mechanisms. It may be:

- **Centralized:** these types of methods utilize a trusted third party or a centralized system to help in authenticating the device.
- **Distributed:** these authentication systems use distributed structure to authenticate devices during any communication. It does not rely on any central server to manage the authenticating information.

Both these categories may utilize hierarchical structure or flat structure. Former method includes multiple level systems for authentication, while later involves no concept of levels.

4. **Hardware-based mechanisms:** In this type of authenticating method, hardware features may be used for the verification process. It may be:
   - **Implicit:** it utilizes the physical features of the integrated hardware system for improving the efficiency of the authentication process. It may include physical un-clonable function or true random number generator.
   - **Explicit:** it means use of external hardware means like trusted platform module (TPM) and a chip to store keys used for verification, to improve authentication process.
5. **Token usage authentication methods:** it is categorized into 2 categories:
   - **Token based:** the authentication methods use some predefined information to serve as a token for authenticating the user.
   - **Non-token based:** each and every time while transferring data, user must enter his credentials for proving its originality.

6. **Authentication process:** it indicates whether it is always required to authenticate both the communicating parties or not. It is dived into:

- **1-way:** it demonstrates the case where it is not mandatory to authenticate both the communicating parties. It means one entity may continue unauthenticated.
- **2-way:** it may be considered as the mutual authentication process in which both communicating parties will prove its identity to each other.
- **3-way:** it represents the case where third part helps in authentication process. Both communicating parties will depend on this third system for mutual authentication.

7. **Deployment time:** it is categorized into pre-deployed authentication procedure or post-deployed authentication procedure based on the time of deployment of authentication method.

2.3 Multi-Agents and Intelligent Environments

Integration of artificial intelligence and Big Data into IoT has revolutionized the lives of human. It has begun an era where everything acts smartly, i.e. system will perform according to users' needs. IoT devices collect data of the surrounding things and the useful patterns are observed by applying Big Data analysis techniques which, in turn, helps in learning user's habits, leading to systematic utilization of devices [16]. To make IoT system more comfortable for human, a multi-agent system has been employed so that IoT devices can act smartly without human intervention. Multi-agent system is the advanced version of centralized system or single-agent system where multiple agents are loosely coupled with each other to complete a common task [17].

'Agents' refers to different entities whether hardware or software which can apprehend the surroundings and may perform accordingly. This system is self-governing and adaptive. Sensing capability of IoT devices helps in accumulating the data related to the environment they work in and pass it to the actuators who can perform accordingly [18]. This system is supported by assisted learning where agents learn from the feedback they receive from users and then develop their knowledge base. After this, when they encounter the same situation again, then they have information of what to do. It may also be considered as adaptive learning. Figure 2.3 shows the integrated IoT system with intelligent system.

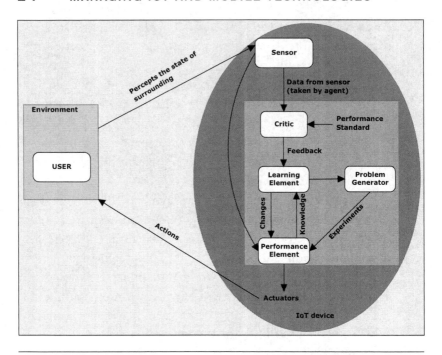

Figure 2.3   Intelligent IoT system.

In this entire scenario, the surrounding information is collected by sensors embedded in IoT devices and presented to the agent for knowledge discovery and withdraws useful patterns for future reference. Here, learning element is added to aid in improving the agent's learning method so as to increase the efficiency of the performance element. Critic assesses the agent's action. It checks whether the actions are appropriate for a given condition or not and results are forwarded to the learning element. Problem generator is responsible for analysing the cases that have not happened earlier. Critic and problem generator are essential elements for effective intelligent system. Performance element receives the input from learning element and decides what to do through actuators and then correspondingly sends the feedback to learning element which can modify its knowledge base. As IoT is a network of distributed devices so multi-agent system is best suited. The key benefits of using multi-agent system are as follows:

- Increases the speed and accuracy because of concurrent execution and asynchronous operations.

- It increases the reliability and robustness of entire system as failure of one agent will not halt the entire system.
- It can cooperate with the growing number of IoT devices as agents may be inserted whenever needed.
- It is a cost-effective system as compared to centralized system.
- Up-gradation of agents is simple and they may be reused wherever required as they are designed in modular fashion.

2.4 Hybrid Cloud and Edge computing

In this section, we describe the notion of cloud computing and edge computing which are pioneering technologies in the IoT paradigm [19–24].

2.4.1 Overview of Cloud Computing

Cloud computing has marked the beginning of novel computing paradigms. It supports various types of clouds that help differently for storage and processing purposes[25]. Figure 2.4 demonstrates the main architecture of cloud computing environment.

2.4.1.1 Public Cloud The cloud infrastructure is made available to large industry group or the public. These clouds are owned and made available by organization selling cloud services. In general,

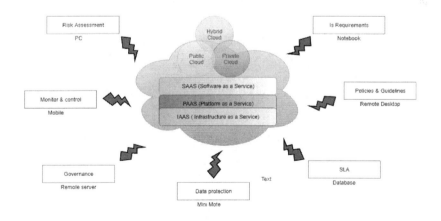

Figure 2.4   Cloud computing architecture.

public clouds offer resources as a service over the Internet connection for a reasonable pay for usage fee. Users are no need to purchase any hardware and can able to scale their use on demand. Generally, these clouds are maintained by a third-party organization that offers cloud service. The public cloud is hosted over the Internet and they are designed such that any user connected over Internet can able to access the services provided by the cloud. Public vendors like Google, Microsoft, and Amazon are offering services to the general public and large industry organizations. All the data created and submitted by the users are generally stored on the servers of the third-party vendors. The following are the advantages of the Public Cloud:

- On-demand Scalability
- Data availability and continuous uptime
- Easy and low-cost setup
- No resources are wasted
- 24/7 (round the clock) technical expertise

2.4.1.2 Private Cloud   This type of cloud infrastructure is operated for only one individual organization. The infrastructure of the cloud can be accessed only by the members of the organization or by the granted third-party members. The sole purpose is not to offer cloud to all the general public, but to use the services within the organization. The private cloud provides more security than the public cloud providers [4,26]. It is more reliable as it supports security and privacy of stored data but incurs higher cost as compared to public cloud.

2.4.1.3 Hybrid Cloud   Hybrid clouds are the most complex model because it involves composition of more than one clouds (like public, Private, Community). Basically, it requires at least one public cloud and one private cloud. A hybrid cloud is generally offered in one of the following two ways:

1. A vendor has a private cloud and makes the partnership with the public cloud service provider.
2. A public cloud provider forms a partnership with the private cloud vendor.

In hybrid cloud, an organization provides some resources from inside and some from outside the house. Hybrid cloud gives the benefits like cost and scale which are available in public cloud and even security and control of the private cloud. Advantages of hybrid cloud are as follows:

- Reduces capital expenses that are part of the organization infrastructure which needs to be outsourced to public cloud providers
- As the public cloud removes the need for investments which indeed improves the resource allocation for temporary projects at a greater reduced cost.
- The hybrid cloud offers both the controls which are available in public cloud and private cloud.
- Cloud-bursting support is supplied.

However, the major drawbacks of hybrid cloud are as follows:

- As the hybrid cloud will extend the IT perimeter outside the organization boundaries, it is going to open up a large amount of surface area for the attack section of the infrastructure which is under the control of service providers.
- When any organizations manage complex hybrid cloud environment using a management tool then either as a third-party tool or a part of cloud platform, the organization should take care of the security implications of the tool being used.
- It is easy to transfer data from the private cloud environment to public cloud in the case of hybrid cloud. There are many integrity and privacy concerns that are associated with such data movement.

2.4.2 Edge Computing

Edge computing [27] refers to the technologies to allow computation which are needed to allow computation at the edge of the network, on upstream of data on behalf of IoT services and on downstream data on behalf of cloud services. Any computing and network resources that are utilized along the path between cloud data centres and data sources is known as an 'edge.' For example, the home things and cloud are connected by the gateways in smart home which is known as edge. The benefits of edge computing are as follows:

- In edge computing, the computing part is done at the proximity of the data sources.

- Face recognition application is improved by moving the computation from cloud to edge the response time is decreased to 900 ms to 169 ms.
- Cloudlets to offload computing tasks for wearable cognitive assistance that improved the response time are between 80 ms and 200 ms.
- Energy consumption is reduced by 30%–40% by cloudlet offloading.

2.5 Legal and Regulations of IoT Privacy and Trust Issues

When we push towards an automated environment, the boundaries between machines and humans have begun to blur. As more and more machines take charge of real-life situations, the issue in the latest roll-out initiative relating to legal and regulatory aspects is becoming very important day by day. Industrial IoT is based on the fundamental principle of gathering a great deal of data on computers, objects around us, processes, and people involved in the process Different components of machines should be able to talk to each other, be able to negotiate, and be able to sign and execute contracts based on the data stream collected and processed [28–30]. When such contracts are not honoured$ and lead to loss in some way, there are so many different questions such as:

1. Is there any human consultant on whose behalf, machines are negotiating contracts with every other?
2. When there are malfunctioning at some stage in the Industrial IOT process, who is accountable and liable in such a scenario? Some suppliers may additionally be concerned in the whole Industrial IoT process, how to decide the perimeter of liability of each provider? The factory administration the place the machines are working, or the design corporations who designed or manufactured the robots, or some different supplier?
3. During the Industrial IoT process, what if the facts accrued by the agency statistics are associated both immediately and indirectly, to individuals, how to defend the privacy problems in such case?
4. A necessary part of the manufacturing method is the supply chain. Supply chain may want to involve integral infrastructures. In quintessential and touchy scenarios, a lot of things in

the furnish chain are managed by using politics, human being's emotions, public opinions, and fast contrast of future reputation. A lot of times, we have a look at the stakeholders go well past the contract to make matters work. In the case of machines taking control, what would occur if the supply-chain problems occur that are hard to provide an explanation for past-agreed contracts.

Here is an attempt to outline associated legal and regulatory factors in the Industrial IoT and the Autonomous world. By no means, this is an exhaustive list; however, it covers huge elements that have to be viewed by means of the key stakeholders while designing and implementation of IoT systems.

- **Fair get entry to connectivity bandwidth:** Net neutrality has always been a warm subject for discussion in many parts of the world. Net neutrality is about dealing with an unfair gain to some companies in pushing their offerings to the quit consumer. As the wide variety of gadgets grows exponentially, the bandwidth demand will outstrip the available supply. If law does not defend net neutrality, there is a top threat that some commercial enterprise may additionally be in a position to take advantage of their muscle groups to manoeuvre the statistics bandwidth wished with the aid of other devices

- **Responsibility, accountability, and liability issues:** As machines are taking decisions and a great autonomy and intelligence are added in the machinery, the questions are being asked about who will be accountable, accountable, and liable for the moves taken on the fly. An apparent instance of this is the self-sustaining car. With all the protection measures in place, there is nonetheless a far-flung possibility of an independent automobile worried in an accident. Now if that occurs – as per law which should be in charge for this accident?

- **Data handling:** Privacy and data safety have been very touchy subjects for the organizations. Many international locations have strict rules about how privacy issues are to be dealt with by means of the organizations. Earlier record access was restricted and was living in closed premises of the groups concerned, and as a result it was easier to deal with troubles and restoration the legal responsibility and responsibility. Now the lines are blurring. In the IoT ecosystem, many agencies are collecting, sharing, and using data with each other. The want is for the businesses to recognize –Is the law of the land allowing the companies to collect, store, and method the data. What is the accountability of the entity storing the facts? Most of the nations have specific laws about whether or what sort of records can be collected, what ought to be stored and how the possession of the records need to be established.

- **Unlawful profiling:** Raw facts being collected with the aid of the devices associated to states and behaviour may want to lead to a correct estimate of the elements like race, age, IQ, personality, political views, etc. The upward push of artificial talent and computing device mastering has led to unique personalization in most of the cases. Organizations ought to preserve in mind how to prevent such profiling that is now not allowed by the law of the land.

- **Privacy and security:** What about the privacy and security of the data underneath use? What are the compliance requirements of the deliberate pilot and full-fledged implementation? Is the end consumer about whom the statistics is being collected, even conscious of the state of affairs and feasible implications of this collection? Is he/she signing for privacy standards? Has the graph taken into considerations of issues like the right to deletion, the right to be forgotten, statistics portability, etc.?

- **Ownership of data:** As units are exchanging the data between themselves and as the heaps of records are getting saved in more than one place, there are many stakeholders and companions worried in the total process. The readability around possession of information wants to be mounted and seemed into very carefully.

- **Automated contracts:** Today, the contracts are performed human to human. Phase of a human-to-human contract is performed and signed between humans. As matters go to a computing device to machine, what takes place if machines begin executing a contract between them? Would machines have the authority to execute such contracts? Can they be a project in the court of law if there is a dispute?

2.6 Future Research Guidelines

In this section, we highlight important future research directions that are still a challenge to achieve. These are given below:

- *Identification methods of infected devices:* currently there is a lack of Internet scale techniques that can identify security issues that exist in IoT devices as many devices are deployed in private environment which restricts its examination and there is not publically accessible dataset for IoT attack signature.
- *Lack of standard security protocols:* as IoT is an evolving field, therefore, various protocols have been designed for its realization. These protocols have their inherent limitations that need to be addressed in future. There is a need of security protocols that can assist in achieving high level of security in IoT implementation.
- *Negligence towards awareness:* one of the key factors behind the infected device passes unnoticed is the immaturity of the IoT device users towards the security of their device. Once installed, they are not maintaining them periodically and correctly.
- *Inappropriate solutions for firmware updates:* there is a lack of solutions that can efficiently patch up the new firmware updates in the devices. Consequently, attacker might exploit the vulnerability exist in older version of software installed in the device.

2.7 Summary

IoT extends networking and communication abilities to the embedded devices with sensing and actuating capabilities. These devices can interact with other devices with less human interference; hence, various security and privacy issues are present in this scenario. Therefore, the main focus of this chapter is to elaborate trust models used in IoT environment to ensure the integrity and correct functionality of the device along with the technologies related to the IoT. It briefly describes various authentication mechanisms employed in IoT network and highlights the legal policies and regulations adopted maintaining IoT security. Lastly, it pinpoints some of the future prospective for the enhancement of IoT security.

References

1. Ziegeldorf, J. H., Morchon, O. G., & Wehrle, K. (2014). Privacy in the Internet of Things: Threats and challenges. *Security and Communication Networks*, 7(12), 2728–2742.
2. Gubbi, J., Buyya, R., Marusic, S., & Palaniswami, M. (2013). Internet of Things (IoT): A vision, architectural elements, and future directions. *Future Generation Computer Systems*, 29(7), 1645–1660.
3. Gupta, B. B., Perez, G. M., Agrawal, D. P., & Gupta, D. (2020). *Handbook of Computer Networks and Cyber Security*. Switzerland: Springer Science and Business Media LLC.
4. Yang, X., & Ding, W. (2019). Researches on data encryption scheme based on CP-ASBE of cloud storage. *International Journal of High Performance Computing and Networking*, 14(2), 219–228.
5. Gupta, B. B. (Ed.). (2018). *Computer and Cyber Security: Principles, Algorithm, Applications, and Perspectives*. London: CRC Press.
6. Gupta, B. B., & Sheng, Q. Z. (Eds.). (2019). *Machine Learning for Computer and Cyber Security: Principle, Algorithms, and Practices*. London: CRC Press.
7. Maher, D. (2017). A human-centric trust model for the Internet of Things. *O'Reilly. Retrieved from: https://www.oreilly.com/learning/a-human-centric-trust-model-for-the-internet-of-things*.
8. Zheng, Q., Wang, X., Khan, M. K., Zhang, W., Gupta, B. B., & Guo, W. (2017). A lightweight authenticated encryption scheme based on chaotic scml for railway cloud service. *IEEE Access*, 6, 711–722.
9. Tewari, A., & Gupta, B. B. (2017). A lightweight mutual authentication protocol based on elliptic curve cryptography for IoT devices. *International Journal of Advanced Intelligence Paradigms*, 9(2–3), 111–121.
10. Alhaidary, M., Rahman, S. M. M., Zakariah, M., Hossain, M. S., Alamri, A., Haque, M. S. M., & Gupta, B. B. (2018). Vulnerability analysis for the authentication protocols in trusted computing platforms and a proposed enhancement of the off pad protocol. *IEEE Access*, 6, 6071–6081.
11. Senarath, A., Arachchilage, N. A. G., & Gupta, B. B. (2017). Security strength indicator in fallback authentication: Nudging users for better answers in secret questions. arXiv preprint arXiv:1701.03229.
12. Devi Priya, K., & Lingamgunta, S. (2020). Multi factor two-way hash-based authentication in cloud computing. *International Journal of Cloud Applications and Computing (IJCAC)*, 10(2), 56–76.
13. Tewari, A., & Gupta, B. B. (2019). A novel ECC-based lightweight authentication protocol for internet of things devices. *International Journal of High Performance Computing and Networking*, 15(1–2), 106–120.
14. El-hajj, M., Fadlallah, A., Chamoun, M., & Serhrouchni, A. (2019). A survey of Internet of Things (IoT) authentication schemes. *Sensors*, 19(5), 1141.
15. El-hajj, M., Chamoun, M., Fadlallah, A., & Serhrouchni, A. (2017, December). *Taxonomy of authentication techniques in Internet of Things (IoT)*. In *2017 IEEE 15th Student Conference on Research and Development (SCOReD)* (pp. 67–71). IEEE.

16. Balaji, P. G., & Srinivasan, D. (2010). An introduction to multi-agent systems. In *Innovations in Multi-agent Systems and Applications-1* (pp. 1–27). Berlin, Heidelberg: Springer.

17. Vlassis, N. (2007). A concise introduction to multiagent systems and distributed artificial intelligence. *Synthesis Lectures on Artificial Intelligence and Machine Learning*, 1(1), 1–71.

18. Boldea, R. (n.d.). The usage of intelligent agents in IoT devices. *Today Software Magazine*. Retrieved from: https://www.todaysoftmag.com/article/2581/the-usage-of-intelligent-agents-in-iot-devices.

19. Yaseen, Q., Aldwairi, M., Jararweh, Y., Al-Ayyoub, M., & Gupta, B. (2018). Collusion attacks mitigation in internet of things: A fog based model. *Multimedia Tools and Applications*, 77(14), 18,249–18,268.

20. Manasrah, A. M., & Gupta, B. B. (2019). An optimized service broker routing policy based on differential evolution algorithm in fog/cloud environment. *Cluster Computing*, 22(1), 1639–1653.

21. Muhammad, G., Alhamid, M. F., Alsulaiman, M., & Gupta, B. (2018). Edge computing with cloud for voice disorder assessment and treatment. *IEEE Communications Magazine*, 56(4), 60–65.

22. Gupta, B. B., Gupta, S., Gangwar, S., Kumar, M., & Meena, P. K. (2015). Cross-site scripting (XSS) abuse and defense: Exploitation on several testing bed environments and its defense. *Journal of Information Privacy and Security*, 11(2), 118–136.

23. Gupta, S., & Gupta, B. B. (2015, May). *PHP-sensor: A prototype method to discover workflow violation and XSS vulnerabilities in PHP web applications.* In *Proceedings of the 12th ACM International Conference on Computing Frontiers*, Italy (pp. 1–8).

24. Bhushan, K., & Gupta, B. B. (2019). Distributed denial of service (DDoS) attack mitigation in software defined network (SDN)-based cloud computing environment. *Journal of Ambient Intelligence and Humanized Computing*, 10(5), 1985–1997.

25. Goyal, S. (2014). Public vs. private vs. hybrid vs. community-cloud computing: A critical review. *International Journal of Computer Network and Information Security*, 6(3), 20.

26. Kaushik, S., & Gandhi, C. (2019). Ensure hierarchal identity based data security in cloud environment. *International Journal of Cloud Applications and Computing (IJCAC)*, 9(4), 21–36.

27. Shi, W., Cao, J., Zhang, Q., Li, Y., & Xu, L. (2016). Edge computing: Vision and challenges. *IEEE Internet of Things Journal*, 3(5), 637–646.

28. Atzori, L., Iera, A., & Morabito, G. (2010). The Internet of Things: A survey. *Computer Networks*, 54(15), 2787–2805.

29. Miorandi, D., Sicari, S., De Pellegrini, F., & Chlamtac, I. (2012). Internet of Things: Vision, applications and research challenges. *Ad hoc Networks*, 10(7), 1497–1516.

30. Roman, R., Zhou, J., & Lopez, J. (2013). On the features and challenges of security and privacy in distributed internet of things. *Computer Networks*, 57(10), 2266–2279.

3

THE MOBILE TECHNOLOGIES IN THE 'INFORMATIVE' SOCIETY

MIAOJIA HUANG, KRIS MY LAW, AND CHENYU XU

Contents

3.1 The Communication Perspective

3.1.1 The Concept of Mobile Telecommunication

Mobile communication refers to communication between mobile users and fixed-point users or between two mobile users. This concept of 'mobile' is not limited to moving people, but can also be moving objects such as moving cars, trains, and ships. Mobile communication technology is designed for wireless communication and it is one of the greatest advancements of mobile Internet and computers. Throughout the past decade, mobile communication is the fastest-growing branch of the communications industry, and has become one of the most important industries at present, and is widely used in all walks of life.

Nowadays, many aspects of human life such as finance, medicine, education, etc., all depend on the use of mobile communication technology. This technology has also changed people's production ways and lifestyles, and even has a profound impact on politics and culture. Thirty years ago, people fantasized about drones, electricity-controlled home appliance, online video chat, online shopping, etc., and these have all been realized. In addition to mobility, mobile communication overcomes the difficulties of electromagnetic wave interference and noise interference, and applies itself to multi-user communication systems and networks, so that different users can work in harmony without interfering with each other.

It is difficult to find another technology that can be so closely related to human life. Understanding the development history of mobile communication and studying its future development trends will help us make better use of this technology and continue to improve people's economic, cultural, and living standards. This chapter reviews the development of mobile communication and highlights the key innovations over the years.

3.1.2 Evolution of Mobile Communication Systems

In 1895, Guglielmo Marconi who is also known as 'the father of radio' used induction coils and load antennas to achieve a 1.5-mile Morse telegraph test in Italy. This experiment also marked the entry of mankind into the era of mobile communication. In 1906, Reginald Fessenden employed the heterodyne method to achieve the first radio broadcast, indirectly promoting the first advancement of radio communication. Then AM signals and FM signals entered the communication system one after another in 1920 and 1933. In 1941 during World War II, radar technology was introduced. In 1946, Bell Labs invented a commercial mobile phone system. Based on this system, the first bus phone network in the world was established in the U.S. This stage is considered as the beginning of modern analogue mobile communications. At this stage, mobile communication only satisfies the information communication needs of a small group of people. From 1950s to 1960s, along with the development of semiconductors and integrated circuits, the advancement of telecommunications technology and information promoted the advent of 'analogue communication era.' Early AMPS experiments based on the concept of a cellular system can only provide voice services in a limited area with small capacity, poor call quality, high cost, and large volume. AMPS is regarded as an earlier 0G improved mobile communication service and is a representative of the first-generation mobile communication service 1G.

Subsequently, the continuous expansion of AMPS system and the advent of cellular mobile communication technology drove the progress of 1G.

In 1991, Europe opened the first 'Global System for Mobile Communication'; thus, opened the 2G era (Meierhellstern et al., 1992). The GSM system began to use SIM cards for user identification, which also means the arrival of the era of digital wireless communications. In July 1993, the CDMA cellular system was successfully developed by the American Qualcomn company and was adopted as the North American digital cellular mobile communication standard (Chih-Lin et al., 1995). 2G services provide more capacity than analogue mobile services and carry more voice traffic. This greatly improves the quality of calls between people in mobile communications.

The IMT-2000 that appeared before and after 2000 marks the world's entry into the 3G era (Medard, 2000). In 2014, 4G has undergone many tests and has begun to be commercialized on a large-scale worldwide. In China, China Mobile, China Telecom, and China Unicom each obtained a TD-LTE license, which marked the official entry of China's telecom industry into the 4G era. By the end of 2010, the mobile phone penetration rate of the global population achieved 70%. The rapid increase in user groups has led to a rapid increase in non-voice services such as images and data, resulting in an increasingly urgent need for wireless communication services and promoting the continuous progress of wireless communication theory and technology (Figure 3.1).

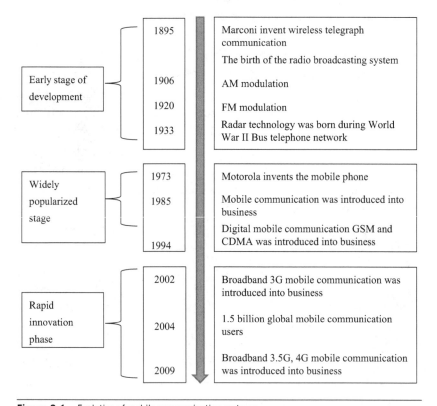

Figure 3.1   Evolution of mobile communication systems.

3.1.3 Introduction of 4G Technology

4G refers to the fourth generation of mobile phone technology. It was launched by Telia Sonera in Finland in 2010, following on from 3G and 2G technology. Nowadays, it has discontinued 3G services in the nation and the 3G customers have now been upgraded to 4G, which has become a mainstream wireless communication network.

3.1.3.1 Key Features of 4G The key features of 4G are faster download speed, lower latency, and crystal clear voice calls.

3.1.3.1.1 Faster Download Speed 4G networks established based on what 3G provides but develops towards a faster speed with the bandwidth of around 200 MB/s, meaning that the maximum theoretical data transfer speed is 200 MB/s. But in practice, you will get the average speed of 25 MB/s, which are around 10 times faster than that of 3G. Imagine that the same film that would have taken you more than 3 hours to download on 3G supposedly takes less than 18 minutes on 4G, depending on your carrier, the location of the cell tower, congestion, etc. Its higher data speeds could make smartphones much more comparable to PCs, giving them better multimedia and gaming capabilities. And the faster speeds meant that mobile gaming was brought to a whole new level.

3.1.3.1.2 Lower Latency In addition to faster download times, 4G has a lower latency, reducing from 80 milliseconds (3G technology) to 45 milliseconds (4G technology). And devices that are connected to a 4G network will react much faster than that of a 3G network. This key feature plays a crucial role when playing online games and streaming live video.

3.1.3.1.3 Crystal Clear Voice Calls Voice over Internet Protocol (VoIP) is similar to Voice over LTE (VoLTE), which is used in WeChat video calls. VoLTE rides on the back of the 4G network, bringing smooth voice calls and high-definition video to 4G phones. In addition, VoLTE allows you to make a call and text at the same time, where previously you aren't able to.

3.1.3.2 Application of 4G   Some typical applications are described below including the smart home, intelligent manufacturing, the Internet of Things (IoT), and entertainments.

3.1.3.2.1 Smart Home   Smart home refers to the use of technical systems, remote-controlled, or other automated devices in houses. 4G technology, as an essential part of error-free and uninterrupted telecom service, accelerates the adoption of home Wi-Fi routers, which is indispensable for the popularity of smart homes.

With the perspective of safety, health, convenience, and energy saving, smart home includes several major subsystems, such as smart video surveillance, smart door locks, smart security, smart home appliances, smart lighting curtains, smart theatre, smart background music, and so on, providing a high-quality lifestyle as a whole.

Imagine these scenarios, turn on the lights when you come, regularize lights not be too dazzling when it's dark, turn off the lights when you are out, turn on the air conditioner when it's hot or cold, turn on the water heater and electric kettle, clean the floor regularly, open the curtains at sunrise, turn off the closed-circuit television when you go home, alarm for abnormalities at home, and finally voice control over all smart appliances. Life is so relaxing and comfortable. Nowadays, above scenarios have become a reality with the development of 4G technology.

3.1.3.2.2 Intelligent Manufacturing   Traditionally, manufacturers rely on wired technology to connect machines and equipment. They use programmable logic controller (PLC) to automate industrial assembly lines, and use field bus to control all devices. Recent years have witnessed that the wireless solutions, including NFC, Ethernet, and IO-Link based on 4G networks, have been used in the manufacturing workshop.

In the framework of Industry 4.0, factory automation increasingly demands more connectivity, flexibility, configurability, and serviceability – all areas where near-field communication (NFC) technology can help provide significant enhancements. In factory automation, data connectivity is essential. A number of standards are used to support specifically developed fieldbuses including the classic fieldbus systems such as Controller Area Network (CAN) as well as real-time

Ethernet. What's more, IO-Link communication networks can be used for bi-directional, point-to-point data connectivity down to the actuator, and sensor level, enabling data pre-processing, sensor parameter tuning, and advanced diagnostics.

3.1.3.2.3 The Internet of Things   The field of the IoT can be summarized as sense, transmit, connect, compute, and control. Specifically, sensors collect data, which are transmitted through the 4G communication module to the cloud server. Then calculate these data and issue commands to the slave machines if necessary. Each phase of the IoT technology is linked to each other, among which 'transmission' depends largely on 4G technology. Thanks to the mature 4G module, those collected data are continuously transmitted to the cloud server.

3.1.3.2.4 Entertainments   First of all, the closest thing to the general public is the home network TV, which is live broadcast through 4G network in real time, supporting 4 K high-definition picture quality and playback.

In the field of live video streaming, the anchors on live streaming platforms such as Douyu and Inke are very active in China, while live game streaming on Twitch is also relatively active in the U.S. As for mobile short video, Douyin and Kuaishou are so popular that people enjoy watching short videos one after another in any place.

With the help of 4G and Wi-Fi, virtual reality (VR) and augmented reality (AR) have stepped into the first stage. At present, VR applications can be divided into computer-side VR and mobile-side VR. Computer-side VR is commonly used in the rendering and modelling of large-scale game computer graphics (CG), while in the mobile-side, with the help of mobile VR professional equipment such as VR glasses, you can watch 360-degree panoramic videos and play VR games. As for AR, 2D AR applications become available, that is, images and texts are superimposed locally, which can be used for operation simulation and guidance, games, remote office, retail, and marketing visualization.

3.1.3.3 Challenges in 4G   Key challenges will be quality of service, security, and complex resource allocation.

3.1.3.3.1 Challenges in Quality of Service   In the future, as the application of 5G continues to deepen, lower latency and higher bandwidth requirements are put forward on the network. Though the throughput of 4G has reached 100 mbps on average, some advanced VR or AR applications require more powerful technical support, such as faster speeds and lower latency. Real-time CG cloud rendering VR/AR requires a network delay of less than 5 ms and a large bandwidth of up to 100 Mbps to 9.4 Gbps.

3.1.3.3.2 Challenges in Security   With the increasing improvement of GPS accuracy, there are pros and cons, and we may be more easily tracked after granting location permissions. What's worse, the risk of misuse of applications will largely increase if devices connecting to Wi-Fi and Hotspot, resulting in interrupting easily in the real scenario (Hassan Gobjuka, 2009). Thus, it is necessary for the telecom operators to build up a security ecosystem among network infrastructure providers and wireless users as a means of protecting data from threats and hackers (Govil & Govil, 2007a).

3.1.3.3.3 Challenges in Complex Resource Allocation   To reduce operation and management costs, devices operating on 4G networks should be able to operate in diverse networks, such as cellular network, wireless LAN, satellite network as well as wireless network. However, accessing these different networks simultaneously is another major challenge for 4G. One probable solution to this problem is named 'multi-mode devices,' which allows end-user devices to get access to multiple networks and services through multi-mode software (Aretz et al., 2001).

3.1.4 Introduction of 5G Technology

To satisfy people's multiple demands for ultra-high traffic density, connection density, and mobility, the development of mobile communication technology and industry has gradually entered the fifth-generation (5G) mobile technology stage (Zhang et al., 2019a). As the latest generation of mobile communication technology, 5G is a complex and a revolutionary progress of the aforementioned four generation technologies. It not only gives full consideration to the diverse connection requirements between people and people, people and objects, objects and objects, but also makes up the defects of the 4G technology in terms of intelligence, flexibility, coverage, security, low energy consumption, etc.

3.1.4.1 Key Features of 5G

3.1.4.1.1 Supportive Access to Massive Devices   5G employs an ultra-dense heterogeneous network and in the near future, it will deploy about 10 times more wireless nodes than existing sites, supporting services for 25,000 users per square kilometre. It not only expands the network coverage area, but also expands the system capacity, which is conducive to improving the network stability.

3.1.4.1.2 High Transmission Capacity and High Data Capacity   The data transmission rate of one user is expected to be 10–20 Gbps, and the capacity of 5G cells can be 1000 times that of 3G and 4G cells (Yang & Zhang, 2015). Different from the traditional single-channel data transmission mode, 5G network is in full operation mode at the same time and frequency, so the processing efficiency and data transmission performance of 5G technology are significantly superior to 4G technology. For instance, the transmission rate increases 10 to 100 times with the peak transmission rate reaches 20 Gbps. The end-to-end delay reaches millisecond level and the connected device density increases 10 to 100 times. In the process of data transmission, 5G can realize timeliness, high reliability and high efficiency, and a high degree of continuity in the downloading process.

3.1.4.1.3 Low-Latency Communication   Latency denotes the time taken for a packet to travel from the sending end to the receiving end. The low latency of 5G enables the state information of the access endpoint to be transmitted to the network in a timely manner, providing guarantee for the intelligent computation of the target and the reliability of the following instructions. 5G network is characterized by ultra-high data rate (>1 Gbps) and low latency at ms level (Yilmaz et al., 2015), which can provide users with high-quality services.

3.1.4.1.4 Reduction of Network Resource Consumption Compared to 4G technology, 5G Network ecology is friendlier. 4G speed consumes a lot of power, especially in the use of mobile phones, computers, and other communication devices, which will affect the performance of these devices. The technical development of 5G takes the network ecology into account and achieves the goal of low energy consumption in each equipment development and application, which will consequently promote the development of a conservation-oriented society.

3.1.4.2 Application of 5G According to the definition of the 3rd Generation Partnership Project (3GPP), 5G technology application includes the following three scenarios (see Figure 3.2): Enhanced Mobile Broadband (eMBB), Massive Machine-type Communications (mMTC), and Ultra-reliable and Low Latency Communications (URLLC).

Among the three application scenarios, EMBB enjoys the transmission rate of 100 Mbps and the peak rate is more than 1 Gbps. Recent years have witnessed the popularity of smart mobile terminals and the real-time transmission of multimedia content has become a trend. The development of 5G enhances the bandwidth of mobile network and the information spread is consequently no longer constrained by the form and content. EMBB is suitable for VR or AR gaming that requires ultra-low latency, streaming 3D 8K videos at 100 fps (Qualcomm, 2020), etc.

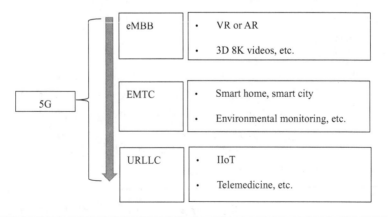

Figure 3.2   Some typical application scenarios of 5G.

EMTC is characterized by wide coverage, low power consumption, and low cost. It is suitable for smart wearable devices, smart home, smart parking, smart city, environmental monitoring, etc. To better achieve the deep integration of industrialization and information technology, the era of the association of everything has emerged.

URLLC communication time is only 1 ms. With the high reliability and fast-moving speed, it has huge market potential and can be applied in many areas such as driveless technology, Industrial Internet of Things (IIoT), telemedicine, etc. With the application demand of driverless trucks, telemedicine, and the proposal of Industry 4.0 concept, low delay and high reliability of data transmission are all required.

Some typical applications are described below including the IIoT, telemedicine, intelligent logistics, and driveless technology.

3.1.4.2.1 Industrial Internet of Things 5G technology with high data transmission quality and stability has promoted the development of IIoT, which focuses on automation, optimization of operational efficiency, etc. A stable and reliable 5G wireless environment can reduce large-scale cabling in the factories, analyse and share real-time data among massive machines and devices (i-Scoop Web, n.d.) and consequently reduce operation and maintenance costs. In addition, due to the complexity of equipment in the workshop, 5G wireless environment enables smooth information communication among massive machine equipment without any human intervention. The larger bandwidth of 5G offers technical support for the data deluge generated in this application (M. Agiwal et al., 2019).

3.1.4.2.2 Driverless Technology 5G technology can realize the transmission of more than 100 IoT devices within 1 square kilometre, and the average transmission rate is around 100 Mbps. It can guarantee the information delay between cars, cars and roads, cars and other things to be within 2 ms, ensuring the safety of unmanned driving. 5G with the features of high efficiency, low delay, and high-density information transmission helps support a unified traffic management centre, which can send the positioning information, navigation information, and environmental information of one vehicle to another without delay and continuously direct the vehicle to complete the whole route.

3.1.4.2.3 Intelligent Logistics   Logistics automation includes transportation, storage, packaging, loading and unloading, distribution, etc. 5G communication technology, IoT, and artificial intelligence (AI) technology are combined to provide data source for enterprise product logistics tracking system through the collection of real-time logistics information. In this way, the distribution routes can be optimized to ensure that goods are delivered to the designated places of customers in the best routes.

3.1.4.2.4 Telemedicine   The goal of telemedicine can also be achieved by the close combination of 5G and the IoT, which will overcome geographical restrictions and become one of the most effective solutions to the medical resources imbalance issues. The medical information stored in the system can be collected, stored, transmitted, processed, and inquired through 5G communication technology to make treatment, diagnosis, real-time health monitoring, health care, or even remote surgery for remote patients.

3.1.4.3 Challenges in 5G

3.1.4.3.1 Challenges in Technical Support   Compared with previous mobile technologies, 5G technology has proposed higher standards in terms of operation rate, equipment capacity, and broadband carrying capacity. Only by upgrading and transforming the traditional transmission network can the communication network transmission effect be optimized, which lays a foundation for 5G network to maintain a good running state. In the development process, 5G technology should be applied to different communication devices and diversified intelligent scenes. Therefore, integration and functionality should be considered to transform the network of different services into a unified network. Meanwhile, deficiencies in the network architecture should be constantly optimized.

3.1.4.3.2 Challenges in Security   Previous generations of networks set up 'protection walls' at the entrance of the network to prevent malicious attacks on the network. 5G network connects multiple devices at the same time; the consequent higher opennness requires 5G to have a more flexible and meticulous security mechanism. All mobile terminals may require security services such as firewalls, intrusion detection systems (IDS), antivirus, integrity checking, and security profiles (Ulltveit-Moe et al., 2011).

3.1.4.3.3 Challenges in Construction Cost Another major challenge for 5G technology is construction costs, including equipment bidding costs, equipment procurement costs, construction costs and base station supporting facilities costs. 5G employs a higher frequency band than 4G, and each base station covers a much smaller area than 4G, resulting in about three times the number of base stations needed to build a 5G network. Therefore, in addition to the investment increase in the early stage, 5G network construction also needs to build more sites.

3.1.4.3.4 Challenges in Operational Management The development of 5G technology involves many industries and application fields, and new operation modes or industries will inevitably be derived in the development process. How to truly realize the development of IoT and how to achieve the cost reduction in the practical application process is another major challenge for 5G. Last but not least, to improve user experience and satisfaction, it is also very tricky to protect and manage user privacy in the process of 5G technology operation.

3.2 The Internet of Things Perspective

3.2.1 What Is the Internet of Things

The IoT is a comprehensive information carrier of the traditional networks and the Internet, which enables the physical entity that have self-identity, self-perception, and intelligence to connect to form a network based on information and communication technology. This network will enable the physical entity to collaborate and interact with each other without human intervention and provide people with wisdom and intensive services. The IoT collects, processes, and monitors the information of the connected items in an intelligent and real-time manner with the support of QR codes, radio frequency identification (RFID), infrared sensors, laser scanners, Global Position System (GPS), and other devices.

It enjoys the features of common object device, autonomous terminal interconnection, and the intelligence of universal services (Tao and Qi, 2017). According to the agreement, any object that connects to the Internet can be used for intelligent identification, real-time location tracking, monitoring, and management (Giuseppe et al., 2017; Andreev et al., 2019; Zhang et al., 2019b). In short, the aim of the IoT is to add computer-based logic to massive things that can be intelligently managed (Daniel & Benedict, 2018).

3.2.1.1 Background of the Internet of Things   The concept of the IoT, also known as Web of Things, was first proposed in 1999 when an RFID was launched at the Massachusetts Institute of Technology's Automatic Identification Laboratory (Auto-IDLabs). The user-experience based innovation is the soul of the development of the IoT (Figure 3.3).

In 2005, at the World Summit on the Information Society (WSIS) held in Tunis, the International Telecommunication Union (ITU) released the report of 'ITU Internet Report in 2005: The Internet of Things,' officially put forward the concept of the IoT. Everything from tires to toothbrushes, houses to tissues can be actively exchanged via the Internet. RFID, sensor technology, nanotechnology, intelligent embedded technology will be more widely used in the background of IoT. Apart from the above-mentioned concept of IoT, characteristics, related supportive

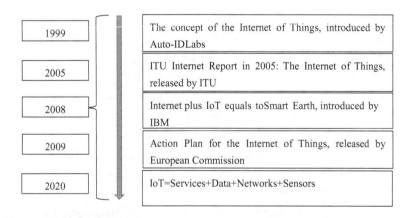

Figure 3.3   Background of the Internet of Things.

technologies, opportunities, challenges, and potential issues of IoT are also introduced in this report.

In 2008, IBM introduced the concept of 'Smart Earth,' that is, 'Internet plus IoT equals to Smart Earth,' and took it as the development strategy of economic revitalization. If the concept of 'wisdom' is embedded in the implementation of infrastructure construction, it can not only stimulate the economy and promote employment, but also build a mature smart infrastructure platform in a short period of time.

In 2009, the European Commission released the 'Action Plan for the Internet of Things,' indicating that it would provide a large amount of financial support at the technical level and put forward a network supervision plan in line with the existing regulations at the government management level.

With the rapid development of the communication technology, especially the rapid development of smart devices, the meanings of modern IoT have changed largely and the applications areas have been widely expanded nowadays, ranging from wearables to agriculture, manufacturing, logistics, etc. (Fernández-Caramés & Fraga-Lamas, 2018). According to the latest conceptual framework for 2020, the IoT can be summarized simply as follows (Atzori & Iera, 2017):

$$IoT = Services + Data + Networks + Sensors \qquad (3.1)$$

3.2.2 Industrial Value of Internet of Things

IoT has offered brand-new possibilities in various topical areas, including agriculture (Gómez-Chabla et al., 2019), industry and engineering (Zaidan & Zaidan, 2020), environment, transportation, logistics, security, medicine (Salagare & Prasad, 2020), and permeates every aspect of our daily life. The aim of IoT is to simplify processes in different application areas, ensure more efficient production processes and system operation, and ultimately simplify our daily life and improve our life quality.

These applications in tourism, service, financial, education, household, and pharmaceutical industry that solidly associated with daily life have been considerably strengthened, which involved all-round of service including different ranges, approaches and value, thus largely enriched individual's life experience. When it comes to defense and military, the impact and the potential value of the IoT would be out-standing even though it is still in its infancy. It can bring enormous value from individual armed forces to the whole facilities to maintain national safety (e.g. aero plane, fighter, submarine, missile, artificial satellite). In conclusion, military accuracy, informatization, and intelligence have been enlarged greatly by the IoT, which enhancing military armed forces, and can be the indispensable factor of the forthcoming direction of military development.

3.2.2.1 *IoT Applications Scenarios*

3.2.2.1.1 *Agriculture Sector*

1. Monitor standardized agricultural production: It collects major critical data factors such as farm chemicals level, soil moisture, and water level in real time to have an exact command of data fluctuation of agriculture yield (Pachayappana et al., 2020).
2. Food traceability: IoT technologies combine vegetable or live-stock with the supply chain. It makes it easier to trace and check the growing environment and production process of the food, to cut down the potentially danger of foodborne diseases (Bai et al., 2017).
3. Hydrological monitoring consists of conventional near-shore contamination testing, ground observation, physical investigation, and satellite remote sensing. It offers an integration platform to collect, transfer, and analyse data for water quality monitoring, which provides experimental and verification means of lake observation and disaster mechanism research.

3.2.2.1.2 *Service Sector*

1. Personal health care: individual can wear certain monitoring sensors to detect their real-time physical conditions. Data collected from the sensors can be transmitted to the health care centre instantly. Once there is any abnormal situation occurred, the health centre would automatically call attention to the carrier to take a medical examination through his telephone.

2. Smart home: it is termed that people can remotely and real-time connect and control the home appliance such as access control, air-conditioning, light, television, and therefore to ensure their safety and satisfy their purpose to enjoy (Rio et al., 2020).

3. Intelligent logistics: modern logistics that empowered by Big Data and IoT can realize a remote vehicle dispatch, automated warehouse management and a coordination among vehicle terminal and dispatch hub (Zhu, 2018).

4. Mobile electronic commerce: IoT enables this business model to achieve mobile payment, buying ticket online, online selling, and so on with mobile devices.

5. Airport intrusion prevention: Sensor strips displaying on the floor surface, fences and other places are placed to impede aggressive incursions like terrorist attacks and smuggling.

3.2.2.1.3 *Public Utilities*

1. Intelligent transportation: Through the GPS and detective system, users can view the operating status of vehicles and show solicitude for expected time of arrival and traffic jam.

2. Security city: Monitoring probes are used to analyse image sensitivity intelligently and connect with 110, 119, 112, etc., in which way builds a harmonious and safe urban living environment.

3. City management: Geocoding technology is used to manage city components by classes and items, which can points out the city management problems.

4. Environmental protection monitoring: All kinds of environmental data collected by traditional sensors are transmitted to the monitoring centre through wireless devices for real-time monitoring and rapid response.

5. Medical and health: This function achieves remote medical treatment, medicine inquiry, health supervision, first aid and visit video surveillance.

It is foreseeable that the further development and maturity of the IoT market will bring huge opportunities and challenges to telecom operators. Furthermore, the saturation of the voice calling market and other factors have also urged telecom operators to turn to IoT market.

3.2.3 *The Analysis of IoT Applications in Logistics Industry*

3.2.3.1 *Industry Sector*

1. Elevator monitoring system: elevators that are equipped with IoT technology can offer various services such as maintenance and data analysis. The system can detect the condition of the elevator such as running speed and aging status. Once an emergency happens, alarms can be transmitted through social media (Zhou, Wang, & Liu, 2018).
2. Monitoring power transmission and distribution and reading meter remotely: Based on the mobile communication network, it realizes real-time collection of power and electricity information, power supply quality data and on-site measurement, facility condition of all power supply points and power receiving points, and remote control of electrical load.

3.2.3.2 *The Impact of IoT in Logistics Industry*   Logistics industry is one of the most practical application fields of IoT-related technologies, which will drastically improve the level of logistics intelligence, informationize, and automation. The further integration of logistics modules will have a positive influence on the operation of every phase of logistics services including production phase, transportation phase, warehousing phase, distribution phase, and sales phase (Sun, K., 2010).

3.2.3.2.1 *Production Phase*   The logistics system powered by IoT can realize the recognition and tracking of raw materials, components, semi-finished products, and finished products on the whole production line. Technically, the Electronic Product Code (EPC) can not only effectively reduce the manual identification costs and error rates, but also quickly select the raw materials and components from a wide range of inventory by electronic tags. Moreover, it can automatically generate detailed replenishment information in advance to maintain a balanced and stable production line.

3.2.3.2.2 Transportation Phase   During the transportation phase, the combination of GPS- and RFID-enabled IoT devices can bring a more intelligent management. By installing various sensors like GPS, temperature sensors on vehicles, they report on everything from real-time location to conditions inside the container and even the remaining fuel levels. By attaching EPC tags on goods and installing RFID sending and receiving devices on some inspection points, workers are able to monitor the quantity and quality of the goods. Besides, we can optimize the intelligent scheduling and arrange the best driving route in order to shorten the transportation time and improve the efficiency.

3.2.3.2.3 Warehousing Phase   Intelligent warehousing based on IoT technology can realize the automatic operation of inventory, stocktaking and pickup, thereby improving the operation efficiency and reduce the operating costs. Real-time stock level monitoring enables planners to know exactly what's on hand and precisely where it is in the warehouse. Therefore, workers can accurately grasp the inventory situation and decrease inventory stocks by making replenishment in time to reduce storing cost. Moreover, advanced analytics operating behind the software that can plan out the most efficient route for Automated Guided Vehicle (AGV) while collecting an order for shipping.

3.2.3.2.4 Distribution Phase   Similar to transportation phase, EPC and GPS technology can accurately locate the position of storage goods in the distribution phase, thus improve the efficiency and speed up the distribution. The picking list can be checked with EPC labels so as to improve the picking accuracy. This function can also reflect other information like the exact number of transhipment in transit, the origin, the destination and expected arriving date.

3.2.3.2.5 Sales Phase   The intelligent shelf will automatically recognize and report to the system through the network when goods with EPC tags are picked up by the customers. In this case, the logistics companies can respond immediately. The companies can also forecast the logistics and service needs according to the historical records so as to carry out active marketing services.

3.2.4 Problems in the Application of the Internet of Things in Logistics Industry

Although the IoT will bring many positive effects to the logistics industry, the application of the IoT is still in a preliminary stage. There are still gaps from our expectations and these problems will be discussed in the following aspects.

3.2.4.1 Technical Aspects   Logistics industry is one of the most complicated industries and there's no doubt that in some fields, the application requirements are beyond the current technical development. Therefore, it is necessary to study the professional domain including information collection, network building and how to process massive sensing information in time and upgrade the original sensing data into knowledge library.

3.2.4.2 Standardization   The existing standards do not come into being a unified one due to the independent implementation of each field in the standard-setting process, lacking sufficient cooperation and coordination among the formulated standards. What's worse, the coding standard is also not unified, causing various difficulties in the technology integration and IoT promotion in the logistics industry. Therefore, the IoT applications in logistics industry urgently require a standard system to achieve the interoperability of item retrieval.

3.2.4.3 Security   As a network platform, the IoT is based on Big Data sharing system, thus its development is inseparable from the protection of data privacy. The current technology and related legal systems have indeed led to a lack of security in the IoT which mainly shows in the following aspects: First, if there is no effective encryption protection measure in the transmission of IoT information data, the data will be easily intercepted or tampered with. The diversity of power saves, battery life, and anti-interference ability in the IoT will all be related to whether the information can be accurately and timely transmitted. Second, the IoT needs to rely on various platforms during its operation, and these supporting platforms should also provide a safe and reliable system for IoT services. Third, because the IoT uses a large number of electronic tags and automatic devices in applications, these devices may threaten data safety and users' information once they are hacked. Fourth, whether the network and routing are stable will also affect the integrity of the data information of the IoT. Therefore, the security of physical equipment is also particularly important. Fifth, general sensors have simple functions and are easily tampered with, which affects their security protection capabilities. Therefore, the firewall strength of sensors is directly related to the security of IoT data.

3.2.4.4 Cost   The second major problem affecting the development of the IoT is cost contradiction. RFID tags can improve customer satisfaction and play an important role in keeping prices low, but RFID tags are too expensive. Manufacturers of general consumer goods need to spend 1.3 million to 2.3 million to implement RFID tags (Hahnel et al., 2004). At present, the cost of RFID tags is about 20 cents, but only when the production volume exceeds 10 billion per year can the cost of tags drop below 10 cents. Although such prices are not worth mentioning for commodities such as cars, refrigerators, TVs, and mobile phones, they are still very high for low-priced products such as light bulbs and toothpaste. For RFID manufacturers, if the cost is too high, the application pressure will be great. If the cost is too low, the manufacturing industry will lose profits again, which is a dilemma.

3.2.5 Analysis of Logistics Industry Development Strategies Based on the Internet of Things

3.2.5.1 Accelerate the Strategic Planning of the Internet of Things Industry and Make It Coordinate with the Logistics Industry Planning   In China, the Yangtze River Delta and other regions have been important places for economic development since ancient times. There are a large number of enterprises and good industrial advantages. At the same time, these areas have ample funds, huge industrial supporting facilities, and the construction appliance for IoT is also sufficient. The success here has driven a wide range of social applications of technology and products, and brought profits to related companies. Therefore, to improve the completion of IoT industry chain, we must fully consider the integrated development of talents, technology, industry and economy, and rely on key cities to vigorously cultivate integrated industrial areas for the IoT.

3.2.5.2 Speed Up Standardization Process   As the country pays great attention on the standardization of the IoT, the government established the China Internet of Things Standards Working Group in 2010, aiming at improving the research and development of IoT technology (Ardissono et al., 2017). This working group is composed of eleven ministries and their subordinates including the Ministry of Industry and Information Technology Working Group on Electronic Label Standards, National Beacon Commission Sensor Network Standard Working Group, National Intelligence Standard Committee, and other 19 relevant standardization organizations. However, the standardize work of IoT involves a huge workload, it requires integrating relevant resources based on the original work and carry out cross-departmental and cross-regional cooperation. The formulation of common technical standards including unified coding rules and standards for middleware interface on basic application platforms should be paid more attention. An open attitude should be placed in the formulation of IoT standards. It is not enough to only develop the IoT of local companies. We should learn more from the experience and lessons of foreign companies. Local companies should be encouraged to establish contacts with overseas companies, share information and materials, thus jointly build a platform for the development of the IoT at home and abroad.

3.2.5.3 Security Strengthen Method   The information safety problem is the most serious threat that RFID technology brings to logistics security. So the technical improvement is required. The label service can be terminated when needed, that is, when the commodity is completed then enter into the transaction phase, an unauthorized reader cannot recognize the RFID label if the information is encrypted or a terminal order is activated. In this way, the related information can be protected. For products that require after-sales service, customer service can be responsible for the decoding and restore the status once the service is terminated. In addition, the access rights of the network should be strictly controlled, except for installing firewalls and anti-virus software. The technical encryption communication channels also need to be improved. Effective laws and regulations should be introduced to punish acts of deliberately stealing the data of others or enterprises, thus ensuring a healthy development environment.

3.2.5.4 Costs Reducing Methods   The IoT can reduce costs by optimizing original data. For example, home appliance manufacturers can remotely log in to the equipment for online debugging which can solve most of the problems of home appliances thus saves the labour cost of maintenance personnel. Through online diagnosis, on-site maintenance personnel can understand the problem and bring the appropriate tools before going to the door to avoid repeated trips. Service subscriptions can increase supplier revenue. For example, a networked water purifier can detect that the filter element is suitable for replacement and automatically order replacement services. Through the combination of multiple companies, consumers can get a one-stop service from thermostats to HVAC. Combining user data with continuous communication can identify products that consumers really want, and maintaining brand interaction can keep consumers informed in time. Collecting data from IoT products will ultimately benefit everyone in the value chain.

References

Agiwal, M., Saxena, N., & Roy. A. (2019). Towards connected living: 5G enabled internet of things (IoT). *IETE Technical Review*, 36 (2), 190–202.

Andreev, S., Petrov, V., Huang, K., Lema, M. A., & Dohler, M. (2019). Dense moving fog for intelligent iot: Key challenges and opportunities. *IEEE Communications Magazine*, 57 (5), 34–41.

Ardissono, L., Gena, C., & Kuflik, T. (2017). *UMAP 2017 PATCH 2017: Personalized access to cultural heritage organizers' welcome. International Conference on User Modeling Adaptation and Personalization.*

Aretz, K. M. Haardt, K., Konhäuser, W., & Mohr, W. (2001). The future of wireless communications beyond the third generation. *International Journal of Computer and Telecommunications Networking (Computer Networks)*, 37, 83–92.

Atzori, L., Iera, A., & Morabito, G. (2017). Understanding the internet of Things: Definition, potentials, and societal role of a fast evolving paradigm. *Ad Hoc Networks*, 56, 122–140.

Bai, H., Zhou, G., Hu, Y., Sun, A., Xu, X., & Liu, X., et al. (2017). Traceability technologies for farm animals and their products in china. *Food Control*, 79, 35–43.

Chih-Lin, I., Pollini, G. P., Ozarow, L., Gitlin, R. D. (1995). Multi-code CDMA wireless personal communications networks. *International Conference on Communication*, Chicago, USA.

Fernández-Caramés, T. M., & Fraga-Lamas, P. (2018). A review on the use of blockchain for the internet of things. *IEEE Access*, 6, 32,979–33,001.

Giuseppe, L., Davide, M., Gabriele, P., Matteo, P., Ovidio, S., & Andrea Azzarà. (2017). An intelligent cooperative visual sensor network for urban mobility. *Sensors*, 17 (11), 2588.

Gómez-Chabla, R., Real-Avilés, K., Morán, C., Grijalva, P., & Recalde, T. (2019). IoT applications in agriculture: A systematic literature review. *ICT for Agriculture and Environment*, 901, 68–76.

Govil, J., & Govil, J. (2007a). *4G mobile communication systems: Turns, trends and transition*. In *2007 International Conference on Convergence Information Technology (ICCIT 2007)*, Gyeongju, South Korea, pp. 13–18. IEEE.

Govil, J. & Govil, J. (2007b) *On the investigation of transactional and interoperability issues between IPv4 and IPv6. IEEE Electronic Information Technology Conference*, Chicago, USA.

Hahnel, D., Burgard, W., Fox, D., Fishkin, K. P., & Philipose, M. (2004). *Mapping and localization with RFID technology. International Conference on Robotics and Automation*, New Orleans, LA.

Medard, M. (2000). The effect upon channel capacity in wireless communications of perfect and imperfect knowledge of the channel. *IEEE Transactions on Information Theory*, 46 (3), 933–946.

Meierhellstern, K. S., Alonso, E., & Oneil, D. R. (1992). *The use of SS7 and GSM to support high density personal communications. International Conference on Communications*, New Brunswick, NJ. Rutgers University, Wireless Inforamation Lab.

Minoli, D., & Occhiogrosso, B. (2018). Blockchain mechanisms for IoT security. *Internet of Things*, 1, 1–13.

Pachayappana, M., Ganeshkumar, C., & Sugundan, N. (2020). Technological implication and its impact in agricultural sector: An IoT based collaboration framework. *Procedia Computer Science*, 171 (2020), 1166–1173.

Qualcomm. (2020). Mission-critical control in 5G – The future of industrial automation. Available: https://www.qualcomm.com/news/onq/2020/02/13/mission-critical-control-5g-future-industrial-automation (accessed April 30, 2020).

Rio, D. D. F. D., Sovacool, B. K., Bergman, N., & Makuch, K. E. (2020). Critically reviewing smart home technology applications and business models in europe. *Energy Policy*, 144, 111,631.

Salagare, S. & Prasad, R. (2020). An overview of internet of dental things: New frontier in advanced dentistry. *Wireless Personal Communications*, 110 (3), 1345–1371.

Scoop. (n.d.). The Industrial Internet of Things (IIoT): The business guide to Industrial IoT, *i-Scoop*. Web. Available:https://www.i-scoop.eu/internet-of-things-guide/industrial-internet-things-iiot-saving-costs-innovation/ (accessed May 8, 2018).

STMicroelectronics. (2009). *Industrial Communication*. Web. Available: https://www.st.com/en/applications/factory-automation/industrial-communication.html#Hassan Gobjuka, "4G Wireless Networks: Opportunities and Challenges."

Sun, K. (2010). A development research of logistics industry based on IOT. *Hebei Enterprise*, (11), 44–45.

Tao, F., & Qi, Q. (2017). New IT driven service-oriented smart manufacturing: Framework and characteristics. *IEEE Transactions on Systems, Man, and Cybernetics: Systems*, 49 (1), 81–91.

Ulltveit-Moe, N., Oleshchuk, V. A., & Køien, G. M. (2011). Location-aware mobile intrusion detection with enhanced privacy in a 5G context. *Wireless Personal Communications*, 57 (3) 317–338.

Yang, L., & Zhang, W. (2015). *Interference Coordination for 5G Cellular Networks*, Cham: Springer, 1–11.

Yilmaz, O. N., Wang, Y.-P. E., Johansson, N. A., Brahmi, N., Ashraf, S. A., & Sachs, J. (2015). *Analysis of ultra-reliable and low-latency 5G communication for a factory automation use case. 2015 IEEE International Conference on Communication Workshop (ICCW)*, London, England,1190–1195.

Zaidan, A., & Zaidan, B. (2020). A review on intelligent process for smart home applications based on IoT: Coherent taxonomy, motivation, open challenges, and recommendations. *Artificial Intelligence Review*, 53 (1), 141–165.

Zhang, W., Kumar, M., & Liu, J. (2019b). Multi-parameter online measurement IoT system based on BP neural network algorithm. *Neural Computing and Applications*, 31 (12), 8147–8155.

Zhang, Y., Deng, J. Y., & Li, M. J. (2019a). A MIMO dielectric resonator antenna with improved isolation for 5G mm-wave applications. *IEEE Antennas and Wireless Propagation Letters*, 18 (4), 747–751.

Zhou, Y., Wang, K., & Liu, H. (2018). An elevator monitoring system based on the internet of things. *Procedia Computer Science*, 131, 541–544.

Zhu, D. (2018). IOT and big data based cooperative logistical delivery scheduling method and cloud robot system. *Future Generation Computer Systems*, 86, 709–715.

4

MOBILE TECHNOLOGIES

SHUANG GENG, YINGSI TAN, AND LU YANG

Contents

4.1 The Business Management Perspective

4.1.1 Mobile Technology Application in Business Management

Mobile technology (Mobile Tech) is mainly composed of mobile applications and mobile devices which provide real-time communication service, data transmission, and sharing (Eng, 2006; Torres et al., 2012). Mobile Tech provides more flexibility for the completion of information-dependent tasks, thus reducing the physical limitation of working places and people (Andriessen & Vartiainen, 2006).

The increased mobility of working places relies on stable and reliable access to the organization's database and document directory, which is further reinforced by the development of mobile technologies and devices (Krcmar, 2010). In modern organizations, mobile devices are playing a more and more important role (Car et al., 2014). Mobile applications and Internet communication technologies are changing the business management methods and how people corporate across organizations. Two main impacts that they have on business operation include the following (Liang et al., 2007):

1. It facilitates communication efficiency among employees, customers, suppliers, and other stakeholders and increases organizational productivity and profitability by taking advantage of enhanced communicating efficiency and information timeliness.
2. It also empowers specific business operations as more alternations are provided for the data access.

Mobile business hardware framework mainly consists of database server, business operation server, Wireless Application Protocol (WAP) or Web Server, firewall, user mobile terminal, laptop and desktop, and so on, as shown in Figure 4.1. Among them, desktops can visit enterprise mobile server applications generally through Internet wired connection while laptops can visit them through wired and wireless connection such as Wi-Fi. In addition, mobile phones also use Global System for Mobile Communication (GSM), Global Packet Ratio Service (GPRS), Code Division Multiple Access (CDMA), Bluetooth, and other wireless communication technologies to visit enterprise mobile server applications (Zhong, 2012).

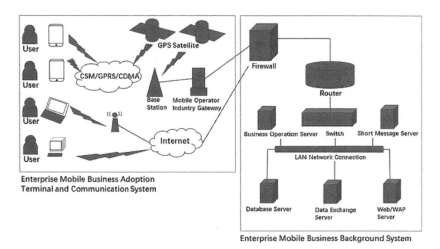

Figure 4.1 Enterprise mobile business hardware framework (Zhong, 2012).

As presented in Figure 4.2, software framework of business management adoptions can be divided into four layers: Mobile Tech layer, system support layer, enterprise application layer, and customer layer. Mobile Tech layer is used for data transmission and accessing. It mainly contains GSM/GPRS/CDMA/Bluetooth, and Internet. System support layer has every kind of mobile terminal operating systems that can interact with enterprise mobile server applications (i.e. Android, Apple OS, Window Phones) as well as operating

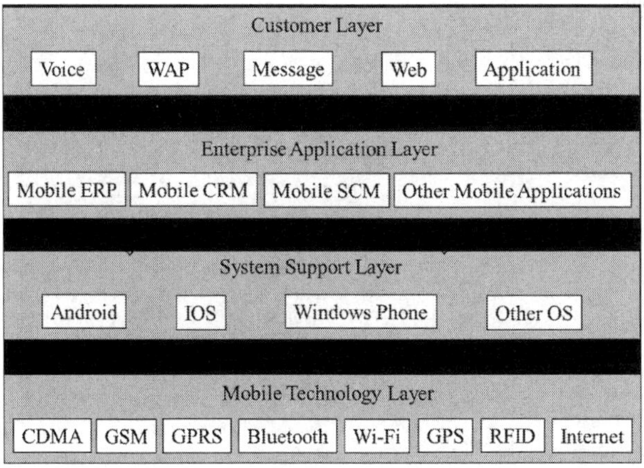

Figure 4.2 Enterprise mobile business management adoptions software framework (Zhong, 2012).

systems like all kinds of laptops and desktops. Enterprise application layer is the server application for mobile Enterprise Resource Planning (M-ERP), mobile Customer Relationship Management (M-CRM), mobile Supply-Chain Management (M-SCM), and other mobile applications. Customer layer refers to the way customers visit such as voice, short message, multimedia message, WAP visit, Web visit, and application client programs.

4.1.2 Characteristics of Mobile Business Management Applications

There are five fundamental features of Mobile Tech, namely, mobility, connectedness, identifiability, interoperability, and personalization (Cousins & Robey, 2015), which accelerate the information sharing and communication, data integration (Mithas et al., 2011), and interoperability across various heterogeneous devices and applications (Cousins & Robey, 2015). These advantages allow managers and employees conduct online discussions and improve the efficiency and effectiveness of decision-making process (Car et al., 2014). Offices and workers located in different regions can take advantage of the time difference and realize the mutual problem-solving frameworks (Koroma et al., 2014).

4.1.3 Mobile Enterprise Resource Planning (M-ERP)

Mobile Enterprise Resource Planning (M-ERP) expands and complements to the traditional ERP system. Its users can use it to transfer, store, and process the latest information to eliminate the delay of business management process. The definition of MERP is that, by connecting the public or private network, the public or private network users can use mobile terminals to connect the ERP system to process the concrete business processes by the means of text, data, voice, and other kinds of information that make mobile terminal compatible and co-working with the ERP system. Through MERP, the user can manage business with higher efficiency, lower cost, and more convenient methods (He et al., 2007).

Homann et al. (2014) analysed the challenges faced by ERP development. The rapid upgrading of mobile operating system programming makes expert feel difficult in finding right programming interface and parameters linked with programming interface. The dynamic market needs also require users to continue to learn new technologies, frameworks, and operation techniques. Standardization is also an important issue for ERP systems. The protocol used by SAP and ERP systems may also need to support Simple Object Access Protocol (SOAP) and Representational State Transfer (RESTful) Web services.

4.1.4 Mobile Customer Relationship Management (M-CRM)

Customer relationship management (CRM) is a critical issue for brand managers and marketers. Technological innovations have great potential to enhance the quality of interaction with customers, thus improving customer managers' working efficiency (Li & Mao, 2012). Mobile Customer Relationship Management (M-CRM) can take multiple forms. The advances of instant communication techniques enable customer relationship managers to interact with consumers through short messaging service (SMS) and other multimedia messaging service. Compared with face-to-face communication, SMS method has the weakness of communication efficiency and incomplete message due to the message length limit.

Mobile users can also access product and service information from Web pages through browsers. This primarily satisfies the information need of consumers, while its interaction capability is relatively poor. Interactivity is an important feature that sustains consumer engagement and brand loyalty. Moreover, the information and system security is also a key issue for business corporations.

With the emergence of smartphone APP and quick database synchronization, more consumers get access to information and communicate with relationship managers more easily. The improvement in modern Mobile Tech also secures the usage of 4G/5G mobile technologies, which enhances user experience (Karjaluoto et al., 2014; Kim et al., 2015).

M-CRM helps to overcome some challenges of traditional CRM, such as the two-way information access between customer (Schierholz et al., 2007; Josiassen et al., 2014; Khodakarami & Chan, 2014) and product or service anytime and anywhere. Additionally, the instant responses from consumers enable brand managers to update marketing strategy and content in real time. These advantages of M-CRM help to achieve higher customer satisfaction (Peltier et al., 2013), thus attracting increased attention from marketers (Kim et al. 2015).

4.1.5 Mobile Supply-Chain Management (M-SCM)

Supply-Chain Management (SCM) is defined as the management of exchanges of materials and information in the logistics process stretching from the purchasing of raw materials to the delivery of end-products to end customers, so linking several firms (Cooper et al., 1997). M-SCM changes the landscape of SCM as it utilizes software and mobile devices to empower managers to conduct SCM activities in a wireless environment anytime and anywhere (Szymczak, 2013). By connecting to the server of corporations, mobile devices enable users to transmit and obtain data across different functions along the supply chain. Mobile supply-chain software application integrates the intra-business and inter-business operations by providing real-time logistics data and requirements. This shortens the time of dealing with customer needs and inquiries, and increases the transparency of the whole supply chain.

From the sustainable SCM perspective, M-SCM strengthens the technological aspect which enhances the stakeholder management and partner collaboration (Seuring, 2013). The transformation from traditional SCM operations to M-SCM requires the integration of multiple corporate factors, for example, the employee development schemes, corporate strategies, management practices, etc. Form a resource-based view, M-SCM reduces the inventory and logistic cost of traditional SCM operations through information exchange and resource optimization. The emergence of Internet of Things (IoT) further changes the corporate business models and promotes flexible design and delivery.

4.2 The Information Sharing at Social Level

4.2.1 An Overview of Mobile Social Network Service

Mobile technologies play an important role in facilitating the building of social connections and relationships (Humphreys, 2008). These technologies, including Wi-Fi, Bluetooth, and Zigbee, have sharply changed our daily lives and now have become indivisible parts of our daily lives. What's more, they enable us to share information at anytime and anywhere for social communication purposes.

Online social networking services (SNS) and mobile social network (MSN) have obtained rising popularity in recent years. They are derived from the people's information sharing and communication needs. In the early Internet environment, E-mail is one of the most widespread social networking tools. Then, the emergence of forums upgrades the 'node-to-node' information exchange mode of e-mail to 'node-to-surface' mode. The continuous advancement of blog, microblog, and other applications facilitates the information transmission and communication between people, and also highlights the individual personalities (Chen, Gao, & Ning, 2017). The rapid growth of usage of SNS persuades SNS vendors to develop mobile applications, such as Facebook app, WeChat, and Twitter, which can be accessed through smart devices (Salehan & Negahban, 2013). The social network over mobile app has a close link with an offline social network, which further increases the popularity of SNS applications.

4.2.2 Development of Mobile SNS

Connecting with others and maintaining social relationship as much as possible is a universal physiological need (Berger, 2014). The emergence of SNS is rooted in the people's need for social interactions and connections. From the early beginning, E-mail communication enables the remote information transmission, which is still a popular communication tool. After that, instant messages and Blog further facilitate the usage of mobile communication technologies and enhance the personalization of information content. With an increasing number of users that utilize mobile technology to connect with others and express personal characteristics in these digital contexts, a social network is formed over these tools, such as Twitter, Instagram, Facebook, WeChat, etc. Mobile SNSs offer some typical and key features that facilitate the online interactions. A user could receive friend invitation and recommendations, and invite others to be connected over the platform. This makes the virtual social network keep growing like a snowball.

The ubiquitous usage of mobile SNS encourages more users to compose and share personalized content for self-expression. The offline daily life is recorded by picture, video, or texts, and shared over SNS. Users spend less efforts in obtaining updates from friends and others, and the virtual SNS becomes kind of reliable information source. Undoubtedly, the shift of ways that people interact also shifts the consumer purchasing habits, e-commerce business models, brand marketing strategies, online education, and other aspects. Especially with the advance of 4G and 5G techniques, users of mobile SNS are keeping exploding. Mobile SNS platforms are competing for more distinctive user experience and stronger business alliances.

4.2.2.1 Mobile SNS and e-Commerce   The mobile SNS has penetrated into our real life in various ways. SNS can connect a wider time and space, and can also change the way people shop. As consumers spend increasing amount of time with social media, social media has become a strategic battle field for marketing campaigns. Different from digital advertisement pushed by Internet browsers, video platforms, and online game providers, advertisements in SNSs allow users to make responses (such as comment, like, dislike, and reply to other's comment) which can signify users' interests and opinions about the ad. This creates potential chances for friends on SNSs to observe others' preferences, initiate discussions, and renew their impressions of the ad or brand. Empowered by this social influence and word of mouth (Li, Lai, & Lin, 2017), advertisement posted on social media might be diffused through the social networks and stimulate users' buying intentions. A bevy of brand managers have sought to engage consumers through social media (De Vries, Gensler, & Leeflang, 2012).

With the help of SNS, merchants assist in the sale of commodities through social interaction, user-generated content, and other means, and apply socialized elements such as communication and discussion to the e-commerce transaction process. The relationship between the store owner and the platform in traditional e-commerce is highly connected, making it difficult for owners to grasp the promotion volume independently. However, in social e-commerce, the store owner is self-contained, relying on his own network to start a business. Consumers can only play the role of consumers in traditional e-commerce. But in social e-commerce, in addition to consumption, they can also become store owners and share high-quality goods to friends to earn platform commissions. The user's role is more flexible and diverse in e-commerce. Moreover, the shop owner can get rid of the shackles of the platform and can promote sales on multiple platforms, greatly improving the circulation of information and driving economic development.

4.2.2.2 Mobile SNS and Education   Since Information and communication technology (ICT) is widely used to support the learning and teaching work in higher education, online learning approach is well developed. Previous studies show that integrating technology into instruction can definitely improve access to information (Ehrmann, 1999). Mobile education allows more people to arrange their own learning plans in a more flexible time without physical constraints. Learners can take portable devices, such as smartphone, to anywhere and get access to their learning content anytime. Different from traditional student learners, public learners can only utilize fragmented time for learning. Many online learning platforms have developed mobile app (i.e. Netease app) for user learning.

Mobile SNSs nowadays have become a multi-functional space for users to share personal related content, including learning interest, skill training content, and individual learning progress. The large-scale coverage of SNS allows people in different regions to share classrooms and discuss together. It breaks the constraints of classroom communication and make idea discussion more efficient. Social interactions in learning groups also enhance the social presence of learners (Garrison, 2011) and create a sense of belonging to community and support freedom of expression. Increased learner social presence shall improve learner engagement in the learning activities (Law, Geng & Li, 2019). On the other hand, given that SNSs also function as a learning and sharing platform, the quality and copyright of learning content is still a challenging issue which may undermine user experience. These pros and cons leave more space for developers, designers, and managers to bring forward new innovations.

4.2.2.3 Other Applications   In the global COVID-19 that broke out in 2020, the National Health Organization of China used Weibo as a platform for the release of important news, which made the information dissemination of emergencies more rapid and transparent. People can learn about different ways to prevent diseases without leaving home. Government agencies publish high-risk areas through Weibo to effectively achieve regional management and prevent the virus from spreading further.

In the working place, people communicate through both formal and informal manners. Disseminating information over the SNS group becomes a normal way in the working scenario. Some companies even develop their own inner-enterprise SNS applications to connect their employees together which may enhance users' commitment to the organization.

For contemporary college students, the emergence of the SNS network provides them with a good dating platform. As universities are promoting an open teaching mode, SNS better meets the social needs of college students and enables college students to meet more people in their learning life as well as accepting more knowledge. Unlike other SNS networks, the authenticity of information on the campus SNS network is often verified and has certain guarantees, so it is safer for college students to learn and socialize through SNS (Li et al., 2017). The scope of the SNS network is large, and college SNS alliances can be launched among colleges and universities to facilitate college students to find like-minded friends and enrich their extracurricular life. At the same time, college students can share their experiences with the help of SNS network, and they can show their personal style. Other college students can learn about the experiences of other friends through SNS and have a certain guiding role in their future life planning.

4.3 The Security Perspective

4.3.1 Security Risk of Mobile Technology

Mobile technology has become pervasive throughout the world and mobile devices get more and more integrated into our normal lives. However, the closer that mobile technologies are connected to us, the more critical are the security threats of mobile technologies. Users transmit their information, both personal and public content, through the mobile applications, thus creating space for attackers or illegal developers to obtain and sell personal information. On the other hand, enterprises provide access gateway for their employees and customers. The leakage of credential business information may cause huge lost for the companies. These threats highlight the importance of technological safety and building trust relationship between the information owners and platforms. In the following sections, the security threats of mobile technologies are introduced based on different functional levels, including application level, web level, and physical level (Choo et al., 2016).

4.3.2 *Application-Level Threats*

Application-level threats have been discussed extensively in existing studies (Faruki et al., 2014). For example, a user may inadvertently grant dangerous permissions while installing a free application (i.e. malware) that may be malicious. Normally, applications allow users to make decision between multiple arcane options during the installation stage. Despite that this gives users choices about their information disclosure, the gap between users' perceptions and the real information access rules will become a trap and burden for users (Becher et al., 2011; Choo et al., 2015; Esfandi & Rahimabadi, 2009; Lane et al., 2011; Pan et al., 2011). Most of the mobile technology users do not know what is mobile badware and unaware of the risk behind it.

Mobile 'badware' can be categorized into spyware, malware, and deceptive ware. Spyware collects users' private information by granting secrete access to devices. Malware aims to control the remote mobile devices by privately insert and run codes on the device. It can also reveal users' data, such as emails and photos, without their acknowledgment or permission (Choo et al., 2016). The malware can be either an application or partially integrated with an application. The developers of malware make profit by frequently posting ads on user devices or directing users to their target websites. When users notice the threat and uninstall the applications, some malwares might be difficult to remove by normal methods. Besides the badwares, there are other activities that threaten mobile security. For example, through the inaudible sound waves, data can be leaked over the Android devices (Do et al., 2015). However, users are uncertain about whether the applications they use now are safe or not, although they know about these threats. Protection of private information cannot be more important today as people tend to perform every task online. It calls for both legal restriction and technological filtering of mobile applications. From the user perspective, avoid installing suspicious applications may also help to reduce the risk.

4.3.3 Web-Level Threats

Mobile web browsers have similar vulnerabilities as the web browsers on the desktops. The most typical risk of web browsers is directing users to malicious target websites where dynamic links are embedded. Through these links, the malware, Trojans, and viruses are spread across the whole Internet. Once the user enters sensitive information (i.e. ID number, bank account information), the attacker will eventually get the personal data and conduct illegal activities.

4.3.3.1 Physical-Level Threats   Besides the intangible risk of information leakage, the property security of mobile devices is also a typical threat. Most of the mobile devices are portable and expensive, making it more attractive for stealers. The loss of devices will not only cause monetary lost, but also exposure key personal data to others. The bank mobile applications may be easily unlocked and result in greater monetary loss. The corporate network can be also hacked through the personal devices (Imgraben et al., 2014; Choo et al., 2015).

4.3.3.2 Privacy Risk of Mobile Technology   Mobile devices (i.e. smartphones) nowadays are designed to be able to sense the user context through various sensors embedded, such as GPS and accelerometers. These sensors are able to collect rich information of users including their physical environments, health information, behavioural habits, etc. Data collected by the sensors is leveraged by applications that focus on different functional areas, such as healthcare (Hicks et al., 2010; Lane et al., 2011), traffic monitoring (Esfandi & Rahimabadi, 2009), and environmental monitoring (Mun et al., 2009). Once users open these applications, they need to grant the access to personal information, such as identity, alias, and location in order to have the application service. Most of the privacy issues can be divided into two types: organizational or national privacy, and personal privacy. Since the governments and enterprises perform activities that related to the public, they store and manage large amount of sensitive information relevant to social development. If the databases are attacked or the data management is not handled properly, the leakage of confidential information will cause huge loss for the public. From the individual perspective, personal information is associated with many personal issues, habits, interest, or needs. Exposing one's personal data to the public will threaten the personal safety and may cause psychological harm to the personal. Therefore, some application developers (i.e. Facebook, Twitter) attempted to take actions for the protection of personal privacy over the mobile devices although it still has a long way to go.

To summarize, security and privacy problems of mobile technology is a critical issue for every users, developers, and service providers. If the trust between users and mobile technologies is undermined, the development of mobile technologies will be hindered. People might choose the simple and old ways for living for the sake of feeling safe.

4.3.4 Safeguard Measures

Considering the importance of security issues when using mobile technologies, researchers and industrial practitioners have dedicated collaborative efforts in protecting users. Encryption is one of the most widely adopted measures to avoid the security threat. Encryption technology generally refers to the encryption and authentication of data and information. The purpose is to keep objects confidential, thus to improve the reliability and security. Encryption techniques usually function in two ways: software-based encryption and hardware-based encryption.

4.3.4.1 Software-Based Encryption   Cipher algorithm is the core of software-based encryption. Despite that many cipher algorithms have been developed over the years, the security of information is indeed guaranteed by the secrecy of encryption keys. Therefore, opening cipher algorithms to the public allows more researchers and attackers to challenge the decrypting tasks, which may on the contrary contribute the improvement of the cipher algorithms.

There are a few widely used software-based encryption algorithms. Rivest Cipher 4 (RC4) is a popular stream cipher whose key length is dynamic and is often used in protocols such as Secure Sockets Layer (SSL). Advanced Encryption Standard (AES) is another encryption method used for wireless communication (Pan et al., 2011). In addition to the stand-alone encryption algorithms, some security measures use strategies to break the private keys into multiple parts and use multiple agents to decrypt. This largely increases the difficulty of malicious attempts to decrypt through mining the private key as each part of the key is allocated in a different node. Another approach is to use an additional private key to protect the original private key (Esfandi & Rahimabadi, 2009).

4.3.4.2 Hardware-Based Encryption   Software-based encryption/decryption is targeted for data transmission security purpose. For the communications between multiple modules in a system, hardware-based encryption is more desirable to bind the modules and devices together. Hardware-based encryption use hardware chips to store the chip information, key information, and hardware information. The encryption relies on the coworking of these three devices. When the main chip generates a random information, the encryption chip will encrypt the information and transmit back to the main chip. If the main chip received expected encrypted information, it will decide to trust the encryption chip and let the controlled operations proceed. In this way, mobile devices need to have the internal knowledge (i.e. encryption algorithm) to be identified by connected devices.

According to the application needs in different fields, researchers have proposed many new approaches for security purposes. In the e-business card sharing scenario (Figure 4.3), one new application was developed which requires each user to perform a gesture before accessing the card information. Malicious attackers cannot access the e-cards as they could not see and repeat the gestures. This helps to secure the e-business card sharing (Liang et al., 2014).

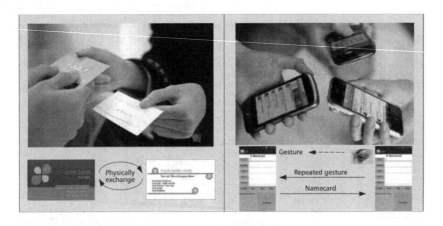

Figure 4.3   Business card application (Liang et al., 2014).

Considering that mobile users are usually not willing to contribute personal information to applications, Li and Cao (2015) advocated that both incentive and privacy aspects should be accounted for in the mobile sensing process. For this reason, they further proposed two credit-based incentives for the sensor systems. One scheme is developed for the case where a trusted third party is available. In this case, the third party is expected to protect user information and defense against attacks. The other scheme deals with the cases where a reliable third party is not available. In this case, blind signature or partially blind signature, and extended Merkle tree are used.

Besides encryption technology, access control technology and data separation technology are also used to ensure security and privacy. Access control technology is a technique that limits the scope, manner of access to data and information resources. By restricting the access rights of users, processes, and services to sensitive resources, the protection of sensitive data and information is ensured, therefore ensuring that the information system is used within the legal scope. When it comes to data separation technology, it uses a distributed system to cut and separate data before storage in which way can guarantee the security of data and information. When data is separated, the data analysis strategy is first combined and the data is cut into data fragments according to the separation threshold. The data fragments can be restored to the original data according to the data separation inverse operation.

4.4 Conclusion

In conclusion, this part firstly outlines the mobile technology application in business side to integrate business systems and extend business competitive advantages. Three key applications, namely, M-ERP, M-CRM, M-SCM are introduced which help to optimize business processes and streamline business operations. In addition, recent years have witnessed SNS or MSN gain increasing popularity in almost all age groups. The rapid development of mobile SNS, such as Facebook, Twitter, Wechat, not only enhance individuals' and organizations' communication efficiency but also highlight individuals' characteristics and personality. Nevertheless, the increasing usage of mobile SNS and other applications allows the collection of user identifiable information. Therefore, it is vital for the users, developers, standard makers, and other stakeholders to be aware of the security and privacy risks. Following this, the countermeasures, such as software-based encryption, hardware-based encryption are discussed. It is not difficult to realize that the technological usefulness and associated challenges or threats always come together. Learning to utilize a particular mobile technology is never the end, yet learning to use it properly and safely is equally important.

References

Andriessen, J. H. E., & Vartiainen, M. (2006). *Mobile Virtual Work – Concepts, Outcomes and Challenges. Mobile Virtual Work.* Berlin Heidelberg: Springer.

Becher, M., Freiling, F. C., Hoffmann, J., Holz, T., Uellenbeck, S., & Wolf, C. (2011). *Mobile security catching up? Revealing the nuts and bolts of the security of mobile devices.* In *2011 IEEE Symposium on Security and Privacy,* Oakland, California USA. IEEE, pp. 96–111.

Berger, J. (2014). Word of mouth and interpersonal communication: A review and directions for future research. *Journal of Consumer Psychology,* 24(4), 586–607.

Car, T., Pilepić, L., & Šimunić, M. (2014). Mobile technologies and supply chain management-lessons for the hospitality industry. *Tourism and Hospitality Management,* 20(2), 207–219.

Chen, Z. K., Gao, J., & Ning, Z. L. (2017). *Mobile Technology and Application* (in Chinese). Beijing, China: Tsinghua University Press.

Choo, K. K. R., Heravi, A., Mani, D., & Mubarak, S. (2015). *Employees' intended information security behaviour in real estate organisations: A Protection Motivation perspective.* In *Twenty-First Americas Conference on Information Systems,* Puerto Rico.

Choo, K. K. R., Rokach, L., & Bettini, C. (2016). Mobile security and privacy: Advances, challenges and future research directions. *Pervasive & Mobile Computing,* 32, 1–2.

Cooper, M., Lambert, D., & Pagh, J. (1997). Supply chain management: More than a new name for logistics. *International Journal of Logistics Management,* 8(1), 1–14.

Cousins, K., & Robey, D. (2015). Managing work-life boundaries with mobile technologies: An interpretive study of mobile work practices. *Information Technology & People,* 28(1), 34–71.

De Vries, L., Gensler, S., & Leeflang, P. S. (2012). Popularity of brand posts on brand fan pages: An investigation of the effects of social media marketing. *Journal of Interactive Marketing,* 26(2), 83–91.

Do, Q., Martini, B., & Choo, K. K. R. (2015). Exfiltrating data from Android devices. *Computers & Security,* 48, 74–91.

Ehrmann, S. C. (1999). Technology in Higher Learning: A Third Revolution Retrieved from: http://www.tltgroup.org/resources/dthierdrev. html.

Eng, T. Y. (2006). Mobile supply chain management: Challenges for implementation. *Technovation*, 26(5–6), 682–686.

Esfandi, A., & Rahimabadi, A. M. (2009). *Mobile agent security in multi agent environments using a multi agent-multi key approach.* In *2009 2nd IEEE International Conference on Computer Science and Information Technology*, Beijing, China. IEEE, pp. 438–442.

Faruki, P., Bharmal, A., Laxmi, V., Ganmoor, V., Gaur, M. S., Conti, M., & Rajarajan, M. (2014). Android security: A survey of issues, malware penetration, and defenses. *IEEE Communications Surveys & Tutorials*, 17(2), 998–1022.

Garrison, D. R. (2011). *E-learning in the 21st Century: A Framework for Research and Practice* (2nd ed.). New York: Routedge.

He, T., Lu, H., & Xu, X. (2007). Research on the key technology of mobile ERP. *Journal of Harbin Institute of Technology*, (4), 8.

Hicks, J., Ramanathan, N., Kim, D., Monibi, M., Selsky, J., Hansen, M., & Estrin, D. (2010). And wellness: An open mobile system for activity and experience sampling. *Wireless Health*, 34–43.

Homann, M., Banova, V., Wittges, H., & Krcmar, H. (2014). Towards an end-user development tool for mobile ERP applications. In *Novel Methods and Technologies for Enterprise Information Systems*. Cham: Springer.

Humphreys, L. (2008). Mobile social networks and social practice: A case study of Dodgeball. *Journal of Computer-Mediated Communication*, 13(1), 341–360.

Imgraben, J., Engelbrecht, A., & Choo, K. K. R. (2014). Always connected, but are smart mobile users getting more security savvy? A survey of smart mobile device users. *Behaviour & Information Technology*, 33(12), 1347–1360.

Josiassen, A., Assaf, A. G., & Cvelbar, L. K. (2014). CRM and the bottom line: Do all CRM dimensions affect firm performance?. *International Journal of Hospitality Management*, 36(1), 130–136.

Karjaluoto, H., Töllinen, A., Pirttiniemi, J., & Jayawardhena, C. (2014). Intention to use mobile customer relationship management systems. *Industrial Management & Data Systems*, 114(6), 966–978.

Khodakarami, F., & Chan, Y. E. (2014). Exploring the role of customer relationship management (CRM) systems in customer knowledge creation. *Information & Management*, 51(1), 27–42.

Kim, C., Lee, I. S., Wang, T., & Mirusmonov, M. (2015). Evaluating effects of mobile CRM on employees' performance. *Industrial Management & Data Systems*, 115(4), 740–764.

Koroma, J., Hyrkkänen, U., & Vartiainen, M. (2014). Looking for people, places and connections: Hindrances when working in multiple locations: A review. *New Technology, Work and Employment*, 29(2), 139–159.

Krcmar, H. (2010). *Informations Management* (5th ed.). Heidelberg: Springer.

Lane, N. D., Mohammod, M., Lin, M., Yang, X., Lu, H., Ali, S., & Campbell, A. (2011). *Bewell: A smartphone application to monitor, model and promote wellbeing.* In *5th International ICST Conference on Pervasive Computing Technologies for Healthcare*. University College Dublin, pp. 23–26.

Law, K. M. Y., Geng, S., & Li, T. (2019). Student enrollment, motivation and learning performance in a blended learning environment: The mediating effects of social, teaching, and cognitive presence. *Computers & Education*, 136 (Jul.), 1–12.

Li, L., & Mao, J. Y. (2012). The effect of CRM use on internal sales management control: An alternative mechanism to realize CRM benefits. *Information & Management*, 49(6), 269–277.

Li, Q., & Cao, G. (2015). Providing privacy-aware incentives in mobile sensing systems. *IEEE Transactions on Mobile Computing*, 15(6), 1485–1498.

Li, Y. M., Lai, C. Y., & Lin, L. F. (2017). A diffusion planning mechanism for social marketing. *Information & Management*, 54(5), 638–650.

Liang, T., Huang, C., Yeh, Y., & Lin, B. (2007). Adoption of mobile technology in business: A fit-viability model. *Industrial Management & Data Systems*, 107(8), 1154–1169.

Liang, X., Zhang, K., Shen, X., & Lin, X. (2014). Security and privacy in mobile social networks: Challenges and solutions. *IEEE Wireless Communications*, 21(1), 33–41.

Mithas, S., Ramasubbu, N., & Sambamurthy, V. (2011). How information management capability influences firm performance. *Mis Quarterly*, 35(1), 237–256.

Mun, M., Reddy, S., Shilton, K., Yau, N., Burke, J., Estrin, D., Hansen, M., Howard, E., West, R., & Boda, P. (2009). *PEIR, the personal environmental impact report, as a platform for participatory sensing systems research.* In *Proceedings of the 7th International Conference on Mobile Systems, Applications, and Services. Association for Computing Machinery*, New York, NY, USA, pp. 55–68.

Pan, J., Qi, N., Xue, B., & Ding, Q. (2011). *Design and hardware implementation of FPGA & chaotic encryption-based wireless transmission system.* In *2011 First International Conference on Instrumentation, Measurement, Computer, Communication and Control*, Beijing, China. IEEE, pp. 691–695.

Peltier, J. W., Zahay, D., & Lehmann, D. R. (2013). Organizational learning and CRM success: A model for linking organizational practices, customer data quality, and performance. *Journal of Interactive Marketing*, 27(1), 1–13.

Salehan, M., & Negahban, A. (2013). Social networking on smartphones: When mobile phones become addictive. *Computers in Human Behavior*, 29(6), 2632–2639.

Schierholz, R., Kolbe, L. M., & Brenner, W. (2007). Mobilizing customer relationship management. *Business Process Management Journal*, 13(6), 830–852.

Seuring, S. (2013). A review of modeling approaches for sustainable supply chain management. *Decision Support Systems*, 54(4), 1513–1520.

Szymczak, M. (2013). Using smartphones in supply chains. *Management*, 17(2), 218–231.

Tieying, L. (2017). The study of Yi Class in the innovative ideological and political education model of colleges and universities under the background of SNS [J]. *Education Research of Shanghai University of Engineering Science*, 56(2), 19–22.

Torres, M. H. C., Haesevoets, R., & Holvoet, T. (2012). *CooS: Coordination support for mobile collaborative applications.* In *Mobiquitous-International Conference on Mobile & Ubiquitous Systems: Computing.* Berlin, Heidelberg: Springer.

Zhong, Y. S. (2012). *Mobile Commerce* (in Chinese). Shanghai, China: Fudan University Press.

5

CLOUD-BASED SMART MANUFACTURING

YC CHAU

Contents

5.1 Introduction

In the Industry 4.0 era, Internet of Things (IoT) devices are everywhere in the manufacturing environment. Organizations use equipment and devices with IoT sensors to collect data and process the data in a real-time base to improve operational effectiveness as well as seek opportunities in new business models. However, deploying these connected devices to get value from the data can be challenging, and one of the proper solutions is the IoT SaaS platforms.

In the 20th century, enterprise resources planning (ERP) was well introduced but yet fully utilized the power of ERP due to the physical limitation of data collection. In the pre IoT era, most of the data were not collected real-time as well as processing in a passive way. In actual practice, it always happens that the physical material flow is running faster than the manufacturing information flow. Even though products ship out from manufacturing site, the corresponding manufacturing information is not conclusive. No immediate actions can be taken due to this delay, and additional loss will be encountered.

5.1.1 Key Enablers in Supporting the Industry IoT

5.1.1.1 ERP – Enterprise Resources Planning   ERP has been introduced to the industry for more than 2 decades; however, there are still quite a lot of pain points encountered by enterprises when using traditional ERP.

5.1.1.1.1 Limited Capacity   When the number of concurrent users reaches the limit, and the system is processing massive amounts of data, network congestion arises, leading to network latency, connectivity issues, and log-in failure.

5.1.1.1.2 Lack of Accessibility   In most cases, traditional ERP can only be accessed under the company's internal local area network, and field and business employees cannot remotely access it. Even if interoperability is achieved through a specific technology, the access speed cannot reach the ideal state, hence lowering the productivity.

5.1.1.1.3 Security Issue   The database is deployed on a server. Once the server hardware or hard disk has a problem, it will quickly lead to data loss or abnormal data access.

5.2 Cloud Computing

Cloud computing was introduced around a decade ago while still a lot of users are not familiar with the cloud computing application.

Cloud is a decentralized computing architecture that can improve network service performance, open data, and inspire new models and new formats by converging five types of resources: network, computing, storage, application, and intelligence. The industrial cloud not only solves the practical problems faced in industrial production but also provides new capabilities for the transformation and development of the industry.

5.2.1 Basic Features of Industrial Cloud

With the rapid advancement of intelligent manufacturing, it is possible to transmit and collect a large amount of production data in real-time, for example, the IoT devices parameters, the critical productivity parameters, and the abnormal quality parameters. How to make better use of data to serve production has become a key task.

Quality control is often implemented in the form of random inspection, but random inspection.

It is difficult to achieve a balance between ensuring product quality and not interfering with the production rhythm. Hence, a multi-factor analysis system has to be used to achieve intelligent, fast, and accurate analysis of factory data, mining key factors, internalizing related model algorithms, and improving internal data analysis capabilities.

Equipment engineers and process engineers combine experience with the manual search and multi-factor manual search methods to find key factors. Usually, the debug process required 6 hours, i.e. investigation and analysis take 3 hours and improvement actions take 3 hours. By applying real-time monitoring in individual process input and output, we are able to detect the abnormality in real time. Single-factor monitoring highly relies on personnel experience to confirm whether the input variance and output variance are correct or not.

If we apply the multi-factor analysis using intelligent analysis (AI, for example), then we are able to detect the abnormality in real time. Analyse the interaction of multiple factors, quickly identify abnormal factors, and help improve production yield. Expedite the interpretation of abnormal cases, quickly analyse the effects of combination factors, find out the root causes, and deal with and prevent them.

5.2.2 The Role of Industrial Cloud

Display Layer

The operation of the industrial cloud system can be monitored and controlled via the master control extensive screen system as the centralized control system and the sub-control system located in each production line. Workers are enabled to review the production status and real-time data collected through 'Workshop Management Kanban' and 'Andon Kanban.' With the incorporation of Kanban, workers are allowed to visualize their workflow, and hence improve cross-functional collaboration and enhance production efficiency. It also enables workers to realize the Work-in-Progress limits; thereby, high-quality products can be produced just-in-time while minimizing the production waste. 3D Visualization System is also being used to improve production planning. By leveraging the virtual reality technologies, manufacturers can design the industrial plant with better production and logistical processes and have a more accurate timing of the whole manufacturing process. Apart from the above display, all of the factory information can be accessed through corresponding mobile applications to enhance the accessibility of the data.

Application Layer

Data in the industrial cloud system will be collected and displayed in real time. Users are also allowed to trace the historical data and review the data analysis report without geographical and time constraints.

Processing Layer

Real-time data will be gathered and processed in the System Data Interface. The industrial cloud system not only processes data concurrently but also provides data storage services, data optimization, and real-time data push, mitigating the risk of network latency, data losses, and other challenges encountered when using ERP.

Acquisition Layer

The transmission of real-time data can be achieved through communication protocols such as RS485, ZTTIBOX (supports more than 300 protocols), IMWATCH, and other mainstream protocols.

Device Layer

The industrial cloud system supports a repertoire of the equipment listed as follows: Mainframe Plus Equipment, Injection Machine, Machining Center, AGV, forklift, Electric spark, wire cutting, Robot (Welding), Tool Setting Instrument, Testing Equipment, Marking Machine, Device layer, Truss Robot, Three Coordinate Measuring Machine, Automatic assembly line, Sheet Metal Processing Unit, Water Treatment System, Heat Treatment Furnace, Engineering Equipment, etc.

5.2.3 The Scope of Industrial Cloud Applications

The industrial cloud technologies can be incorporated in the production of smart products and services, respectively, to enhance their performances. The former examples include CPS (Cyber–Physical System), sensors, and product performance simulations while the examples of the latter include digital twin, status monitoring, sensors, the IoT, virtual reality, and augmented reality. Other than the smart products and services, the technologies of the industrial cloud are widely adopted among smart factories. Operation efficiency and productivity can be ameliorated by integrating the cloud technologies into the process of process simulation, MES, visual inspection, APS (Advance Production System), workshop networking, health management of equipment, additive manufacturing, and data acquisition (SCADA Supervisory Control And Data Acquisition). The technology also contributes to the creation of intelligent equipment, intelligent and collaborative robots, an intelligent production lines that help to uplift the factory operation performance. An in-depth case study and a detailed workflow will be illustrated later in this book chapter. Besides, research and development can be achieved in a 'smart' way by assimilating the cloud technologies into embedded software, design cost management, DFM (Design For Manufacturing) analysis, Topology Optimization, CAS (Computer Aid System), CAM (Computer Aid Manufacturing), EDA (Electronics Design Automation), PLM (Product Life Management), and CAE (Computer Aid Engineering). Apart from the above applications, the industrial cloud system is also ubiquitous in management.

Typical applications include ERP (Enterprise Resources Planning), CRM (Customer Relationship Management), SRM (Supplier Relationship Management), EAM (Enterprise Asset Management), MDM (Mobile Device Management), QM (Quality Management), and EPS (Enterprise Portal Systems). Smart logistic and supply chain management have also leveraged the cloud system in constructing an automated database, DPS (Digital Picking System), TMS (Turnkey Manufacturing System), WMS (Warehouse Management System), AGV(Auto Guided Vehicle), and SLAM (suspended layer additive manufacturing). Last, enterprise decisions can be made in a 'smart' way, optimizing the efficacy through employing BI, industrial Big Data, EPM, and mobile applications in the decision-making process.

5.2.4 *Three Modes of Industrial Cloud*

Smart + Product

By incorporating cloud infrastructure and embedding algorithms into the hardware systems, users can visualize the cloud technologies in the product. The transformation process allows artificial intelligence to become a product function and a unique selling point of intelligent products.

Smart + Service

Instead of selling hardware, this mode allows the intelligent software to be delivered as a service. The SaaS can be licensed and distributed on a subscription basis.

Smart + Production

This mode is also known as production intelligence or manu-facturing intelligence. Production data extracted from various sources can be gathered and analysed to facilitate decision-making process (i.e. the forecasted demand, production run time, etc.), and allow users to visualize the entire production in the dashboard inside the factory.

5.3 Why Should Enterprises 'Go to the Cloud'?

5.3.1 *Flexibility*

The flexibility of traditional ERP is low since its set up does not support a high degree of customization. Owing to the rapidly changing market environment, a flexible ERP is necessary to support the process change in an attempt to satisfy changing demands. Therefore, the 'go-to cloud' is essential to enable enterprises to be responsive to the continually changing environment and manage their operation effectively.

5.3.2 Scalability

Since the capacity of traditional ERP is limited, it can only handle a limited amount of data concurrently, and it is only applicable for small and medium-sized enterprises to use. In contrast, 'Going to the cloud' supports massive data processing with rapid speed. Its unlimited capacity can support and provide services for enterprises on any scale.

5.3.3 Liquidity

As users are only allowed to operate the traditional ERP system connected with the internal LAN, remote access is not prohibited. 'Going to the cloud' is a perfect solution for enterprises to optimize their LAN. The provision of high accessibility also grants users distanced access to the system through applications in portable devices, either via intranet or extranet browsers.

5.3.4 Cost

Implementing traditional ERP software is extraordinarily expensive as it involves setting up & installation costs, maintenance cost, training expenses, impairment cost, purchase costs of computer hardware, and security software. Frequent software upgrades are also needed as the company expands. On the contrary, users are only obligated to pay the periodic subscription fees to cloud service suppliers after 'going to the cloud,' and the supplier will provide a range of after-sales services when necessary.

5.3.5 Security

Due to the limitations of traditional ERP systems such as connectivity issues and capacity problems, data transfer and data storage will be detrimentally affected, leading to lost data and failure to access some of the data. Cloud systems will mitigate the above risks through implementing security safeguard architecture. Various measures, such as preventive controls and detective controls, are enforced to optimize the protection of data and minimize all the potential risks.

5.4 What Is Cloud Computing?

Cloud computing is a type of distributed computing, which involves the process of breaking down the data calculation (Big Data) processing program into infinite small programs, analysing the small programs via a system composed of various servers, and sending the analysed result back to the users. However, multiple challenges are faced by Big Data, hence hindering its analysis performance.

5.5 Massive Data Challenges

As data are growing and accumulating expeditiously over time, the massive volume of data collected may overwhelm the cloud computing system. Therefore, the cloud computing system faces the challenge in maintaining its stability in processing the massive data as well as the accuracy in computation. Besides, data fragmentation occurs when massive data collected are broken down into small pieces. Since data are collected from numerous businesses with complex structure and characteristics, different underlying systems and other multiterminal such as PC, wireless, OTT, and IOT, it is formidably difficult for the cloud computing system to locate and gather the correlated data. Moreover, massive data collected are existing in different structures and different business standards, leading to the rising difficulties in standardizing the data and come up with a fair analysis. Ultimately, the sources and quality of data collected are uncertain. The existence of dirty data, including duplicated data, insecure data, and inaccurate data, etc., may deteriorate the accuracy of analysis results.

5.6 Cloud Computing Interactive Mode

Infrastructure as a Service (IaaS)
Infrastructure as a Service (IaaS) is a cloud infrastructure service that allows users to access various computing resources

over the Internet, which mainly targets enterprises and operators. Users can run the application on the server that they have rented. Examples of IaaS providers include Linode, Microsoft Azure, Amazon Web Services (AWS), etc.

Platform as a Service (PaaS)

Platform as a Service (PaaS) is a cloud service platform provided by independent resource developers that are mainly used for application development. Paas delivers a platform for businesses to design and create customized applications online with special software features. A range of value-added features provided includes data analysis, artificial intelligence, docker, communication, voice recognition, statistics, advertising, etc. Users can concentrate on designing the software without concerning any problems related to the infrastructure, operating system, storage, etc. Examples of providers are ranging from Google App Engine, OpenShift, Window Azures, etc.

Software as a Service (SaaS)

Software as a Service (SaaS) is known as software application services that are commonly designed for both individual users and business. Since the software application is delivered online and managed by a third-party vendor, no installations and downloads are needed. Typical providers include Google GSuite, Salesforces, Dropbox, etc.

5.7 Cloud Service Configuration Mode

Public Cloud

Public cloud is a cloud computing service provided by third-parties publicly over the Internet. It is available to the public who are willing to pay for the resources according to their demand. Customers do not have to worry about the installation maintenance issue as the cloud service provider is obligated to manage the system. Nevertheless, the security problem of public cloud has been raised, and the price is cost-efficient only when a small amount of data and resources are subscribed. Typical industries that are adaptable to public cloud are education, video broadcasting and gaming.

Private Cloud

Unlike a public cloud, private cloud computing services are internally offered to designated users only while the general public are not eligible to access. Although the installation cost is relatively higher than that of public cloud, private cloud is more scalable and flexible as it provides customization services with a better control system. The security level is also higher to ensure the confidentiality of private clouds and restrict the public's access. Industries such as manufacturing, corporate management, government agencies, and aerospace are likely to adapt to private clouds.

Hybrid Cloud

Hybrid cloud is a cloud computing service that consists of mixed features of public cloud and private cloud. Users are enabled to enjoy the benefits of both types of clouds and granted greater flexibility as they can change the cloud platforms freely when necessary. However, the setup procedures of hybrid clouds are more complex, which leads to maintenance difficulties, security challenges, as well as low compatibility. Nevertheless, industries such as finance, healthcare, and agriculture are used to utilize hybrid clouds to support its operation owing to the fluctuating market demand and other constantly changing factors.

References

Chau, Y. C. (2016). Engineer: Cover story – Industry 4.0 with implications for Made-in-China 2025 (Vol 44), May Edition, *Hong Kong Institution of Engineers*. http://www.hkengineer.org.hk/issue/vol44-may2016/cover_story/

Hu, B. P. (2020). White Paper – Industrial Internet of Things Platform, *WSI Operation Handbook*. http://www.wsinnti.com/html/jszl_1704_2845.html

Inauen, C., & Mann, J. (2019). White Paper – Unlock the Power of Industrial IoT with Analytics, *Siemens Digital Industries Software*. https://images.assetsmanu.endeavorb2b.com/Web/PentonManu/%7Be7ca79de-f3fd-4401-afea-4dc0652e74ed%7D_Unlock_the_power_of_Industrial_IoT_with_analytics_(2).pdf

Ranger, S. (2018). What is Cloud Computing? *zdnet.com*. https://www.zdnet.com/article/what-is-cloud-computing-everything-you-need-to-know-about-the-cloud/

6

Cloud-Based SMART Manufacturing – An Implementation Case in Printed Circuit Board Assembly (PCBA) Process Optimization

VINCENT WC FUNG

Contents

6.1 The Background

The case company is a very successful Hong-Kong-based manufacturer with business in OEM and ODM (Original Equipment Manufacturer and Original Design Manufacturer) of sophisticated electrical and electronic products with its factory located in the Pearl

River Delta, Guangdong province, China. Due to substantial growth in electronic items, strict and complicated product design, such as mini mobile phone and micro IoT sensor devices, the case company was motivated for quality improvement and productivity performance in the electronic production workflow. It appears that approximately 75% of the quality defects are related to the solder paste printing process which is the very frontend operation in the printed circuit board assembly (PCBA) production line. Figure 6.1 shows the set-up of the current PCBA process in the electronic industry.

The existing failure detection by Solder Paste Inspection (SPI) and Automated Optical Inspection (AOI) in the PCBA production flow to locate the failures are reactive [1,2], which may create waste and require additional effort to be spent re-manufacturing and inspecting the PCBs. Such process performance of the assembly process cannot be guaranteed at a high level in the rapid growth of complicated electronic products, quality, and the productivity requirement.

Two significant challenges exhibit in a typical PCBA process which offers room for improvement. First, the parameter settings of the PCBA machines are now set by the domain experts, such that the

Figure 6.1 Existing PCBA process.

entire process is highly reliant on their expert knowledge levels and knowledge on the identification of critical factors cannot be transferred to others effectively. Lacking an effective way for knowledge retention and transfer complicates the PCBA process and makes improving the process performance difficult. Second, the existing inspection process is used to investigate the yield loss in the entire PCBA process without considering the effect on machines' parameter settings. Thus, optimization of the process performance is challenging to achieve, and the yield performance solely relies on the domain experts. Figure 6.1 shows the existing PCBA production process which is most commonly using in the electronic industry.

6.2 The Adoption of Intelligent Manufacturing Process Improvement System (IMPIS)

To address the challenges in the existing PCBA process, an intelligent manufacturing process improvement system (IMPIS) is suggested as shown in Figure 6.2. Three modules in the IMPIS for identifying the

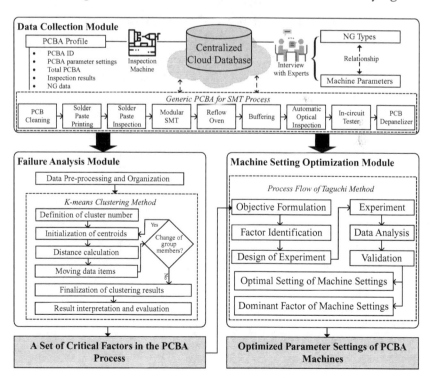

Figure 6.2 The system architecture of the IMPIS.

key factors in PCBA and optimizing the machines' parameter settings are (i) a data collection module (DCM); (ii) a failure analysis module (FAM); and (iii) a machine setting optimization module (MSOM).

6.2.1 Data Collection Module (DCM)

Two types of data are aggregated from the inspection machines and domain experts. First, from the inspection machines, such as SPI, the dataset related to the manufacturing performance can be obtained to establish the PCBA profile. It contains five essential forms of data for the analysis, including ID, the machine's parameter settings, the total number of PCBAs, inspection results, and NG records.

As shown in Figure 6.3, the inspection results in the PCBA process can be divided into four major types, namely 'Good', 'Warn', 'Confirm', and 'NG'. The 'Good' and 'Warn' results indicate that the semi/finished products are (i) within the specification and tolerance and (ii) within the specifications but out of tolerance, respectively, in which both are acceptable, and the products pass the manufacturing process. The result 'Confirm' indicates that the inspection machines themselves cannot determine the PCB situation, and the inspection results need to be approved by the experts in the production line. The result 'NG' refers to the PCBs which are out of specifications; such PCBs are then classified as having failed the manufacturing process. As such, a dataset of the performance of the manufacturing process can be obtained.

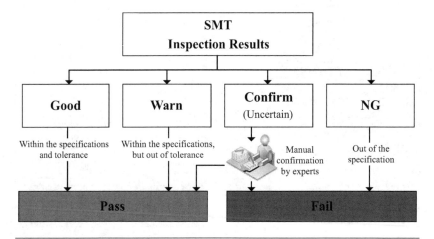

Figure 6.3   Graphical illustration of the PCBA inspection results.

The above-collected data from the inspection machines, such as SPI and AOI, are transmitted to the centralized cloud database (CCD) for the subsequent data analytics, in which the inspection machines are connected to the edge devices and computers in the PCBA process. Thus, the data from various production lines can be gathered to perform the analysis and investigation.

Regarding the number of NG types, the interview with the domain experts is required to investigate the relationship between NG types and machine parameters. In other words, the relationship matrix can be formulated such that the NG problems can be alleviated by considering the specific machine parameters. All of the above data are then stored in the CCD for effective data query and analytics. Since it is difficult and inefficient to reduce all NG types in the PCBA process, Big Data analytics is applied to identify the severe NG types generated by the process.

6.2.2 Failure Analysis Module (FAM)

To classify the severe and non-serious NG types, the k-means clustering method is applied. The historical data of the PCBA profile are considered, in which a matrix $N = (n_{ij})$ is formulated where $i \in n$ denotes the number of PCBAs with the maximum number n and $j \in m$ denotes the number of NG types with the maximum number m. Before using the k-means clustering, several dimensions $D = \{d_1, d_2, ..., d_k\}$ to analyse the historical data are defined by organizing and modifying the data in the matrix N, where the value o is the maximum number of the defined dimensions. Accordingly, the NG types can be analysed according to their features, such as frequency and average occurrence, and the NG types are given with the corresponding coordinates $x_j = \{x_{j1}, x_{j2}, ..., x_{jk}\}, j \in m$ in the k-dimensional plane. In terms of the process of k-means clustering, a fixed number of centres $c = \{c_1, c_2, ..., c_p\}$ should be identified, where p is the maximum number of centres. At first, the number p of the NG types x_j is selected randomly to perform the role of cluster focal points, such as $x_1 = c_1$. The Euclidean distance between the NG types and centres are then calculated, as in Equation (6.1). Subsequently, the NG types are assigned to the clusters based on the minimum Euclidean distance and the new cluster focal points are calculated by averaging the member values of the clusters. The Euclidean distance between the NG types and new cluster focal points are then re-evaluated. The above process is repeated until no movement of the cluster members is observed. The NG types can therefore be classified into p clusters, and the most severe members can be identified:

$$d_{jq} = \left\| x_j - c_q \right\|, \forall q \in p \tag{6.1}$$

Afterwards, the relationship matrix between NG types and machine parameters can be applied to investigate the most critical factors in the PCBA process. A set of critical factors $F = \{f_1, f_2, ..., f_s\}$ can be identified to conduct the experimental study, where s is the maximum number of factors.

6.2.3 Machine Setting Optimization Module (MSOM)

By taking into account a fractional factorial experimental design, i.e. an orthogonal array, to reduce the number of experiments, it is considered to be robust. In the experimental design, there are three types of quality characteristics of response y to be considered when measuring the signal-to-noise ratio (η): (i) larger is better; (ii) smaller is better; and (iii) nominal is best, as in Equation (6.2). The objectives, or responses, are defined at the beginning, and the critical factors and their levels are organized.

After the experiments are conducted per the orthogonal array, the signal-to-noise ratio η_{ijk} can be obtained using the typical Taguchi method, where i denotes the number of factors, j denotes the number of factor levels, and k denotes the number of responses in the experimental design. For the repeated experimental trials, the average signal-to-noise ratio $\overline{\eta}_{ijk}$ can be obtained. Subsequently, the weight of the signal-to-noise in each response w_{ijk} can be calculated using min-max normalization, as in Equation (6.3).

By averaging the weight of signal-to-noise among the responses, the finalized weights can be obtained to identify the best factor setting in the experimental study. Accordingly, the optimal machine parameter settings can be evaluated using the above mechanism, in which multiple objectives can be effectively considered:

$$\eta = \begin{cases} -10 \cdot \log\left(\dfrac{1}{n}\sum\dfrac{1}{y^2}\right), \text{ for "larger} - \text{is} - \text{better"} \\ -10 \cdot \log\left(\dfrac{1}{n}\sum y^2\right), \text{ for "smaller} - \text{is} - \text{better"} \\ \quad 10 \cdot \log\left(\dfrac{\mu_y^2}{\sigma_y^2}\right), \text{ for "nominal} - \text{is} - \text{best"} \end{cases} \quad (6.2)$$

$$w_{ijk} = \frac{\overline{\eta_{ijk}} - \min_{k}\left(\overline{\eta_{ijk}}\right)}{\max_{k}\left(\overline{\eta_{ijk}}\right) - \min_{k}\left(\overline{\eta_{ijk}}\right)} \quad (6.3)$$

6.3 Implementation of the IMPIS

Figure 6.4 shows the real-life SP and next is SPI machine in the PCBA operation. The entire implementation of IMPIS was divided into three phases.

Figure 6.4 Real-life SP and SPI machines in the PCBA.

6.3.1 Phase One: Structured Data Acquisition by DCM

First, there were two major types of data to be collected, i.e. (i) the historical PCBA performance report and (ii) the relationship between factors and NG types. The performance report summarized 84 PCBA samples with 658,179 handled PCBs in total. In the historical records, the PS machine's parameter settings solely relied on the expert and followed the standard industrial practice. The NG results were obtained concerning the PCBA. There were 14 types of NG considered in this study, including no solder (NS), height low (HL), height high (HH), volume low (VL), volume high (VH), area low (AL), area high (AH), offset x– (OX–), offset x+ (OX+), offset y– (OY–), offset y+ (OY+), bridge (BR), shape (SH) and splash (SP).

Second, by interviewing with the domain experts in the electronics manufacturing industry, the relationship matrix was formulated. It shows the correlation between the 14 mentioned NG types and five controllable factors, which were pressure, speed, blade angle, cleaning cycle, and the separation distance between the stencil and PCB.

6.3.2 Phase Two: Pressure, Speed, and Blade Angle (PSA) Model Formulation by FAM

Based on the collected data, the FAM can be applied to analyse the historical data, and the results can be interpreted using the relationship matrix. To analyse the historical PCBA performance records, three dimensions were used, namely: (i) the average number of NG results; (ii) the min-max difference of NG; and (iii) the occurrence of NG. Since the defect rates from the solder paste printing process collected from the SPI are merely the frequency of NG defects, it is difficult to determine the seriousness of the NG defects systematically.

Traditionally, setting a threshold to inspect the NG defects is relatively ineffective due to the subjective judgment on the threshold values. To determine the serious NG types in the solder paste printing process, three above features are considered from the NG defects. The first feature shows the average frequency of the particular NG defects to be observed in the solder paste printing process; the second feature refers to the range in term of the min–max difference of NG defects to understand the variation of the severity; the third feature presents the summarized data collected from the SPI in term of the NG defect types. Provided that the above three features are obtained, the k-means clustering can be applied to categorize the types of NG defects. The historical data were then converted to the dataset to be evaluated by k-means clustering. Subsequently, by using the FAM, it was found that the types NS, HL, HH, VL, VH were classified in cluster 1 (serious factors) and the types AL, AH, OX–, OX+, OY–, OY+, BR, SH, and SP were classified in cluster 2 (non-serious factors). The results from the FAM show that the NS, HL, HH, VL, and VH types of NG were serious and frequently occurred in the existing SP process.

Based on the finding, the pressure (P), speed (S), and blade angle (A) were commonly implicated in the severe NG types, which were regarded as the critical factors in the SP process. Consequently, a pressure, speed, and blade angle (PSA) model for experimental studies was formulated to identify the best parameter settings in the SP process to optimize the process performance.

6.3.3 Phase Three: Optimal SP Machine's Setting by MSOM

By applying the orthogonal array and the experimental data to minimize the number of 'NG' and 'Confirm' results can reduce the defects and manual inspection time, respectively, and thus the process performance can be further improved.

According to the FAM, the calculated signal-to-noise ratios were then averaged to obtain the summarized signal-to-noise ratio in terms of 'NG' and 'Confirm' responses. The results with respect to 'NG' and 'Confirm' were obtained using the Taguchi method in the format of the signal-to-noise ratio. Subsequently, the signal-to-noise ratios were normalized between 0 and 1 as the weighting in each combination. By averaging the normalized weights, the optimal result was obtained.

6.4 Comparison before and after the Implementation of IMPIS

In the traditional SP process, there is no pre-production investigation, and the domain expert is responsible for tuning the machine's parameter settings and maintaining the required process performance.

According to the dataset of historical SPI performance used in the DCM, the average yield loss from 84 PCBA samples with 658,179 handled PCBs was approximately 2.265%.

After the IMPIS is used, a set of experiments needs to be conducted, which uses $9n$ number of PCBs, where n is the number of trials in the experimental studies. The average yield loss after using the proposed system became approximately 0.126% in the SP process.

As such, the waste from the traditional approach can be modelled as no. of NG = 2.265 % × Total PCB, while the waste from the proposed approach can be modelled as no. of NG = 0.126 % × Total PCB + $9n$, as shown in Figure 6.5.

The management of the PCBA process should decide whether to use the proposed approach, depending on the number of PCBs to be handled. Even though increasing the number of trials considered in the MSOM can improve the reliability of the results, the waste and resources for the experimental studies increase as well. For instance, when considering the three trials in the use of the MSOM, the break-even point to select the adoption of the proposed system is at 1198.72 PCBs to be handled. Thus, in this situation, when the manufacturing process must handle more than 1,199 PCBs, the use of the proposed system can be justified and can outperform the existing methods.

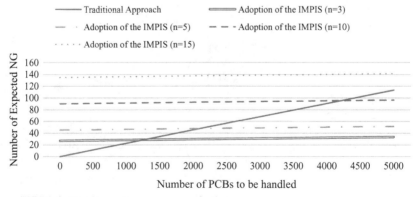

Wastes from Traditional and Proposed Approaches

——— Traditional Approach   ═══════ Adoption of the IMPIS (n=3)

— · Adoption of the IMPIS (n=5)   — — - Adoption of the IMPIS (n=10)

· · · · · Adoption of the IMPIS (n=15)

Figure 6.5   Break-even point evaluation before and after implementing IMPIS.

6.5 Exploration of the Dominant Factor in the SP Process

From the IMPIS results, it was found that the factor of blade angle (A) was dominant in the entire SP process, in which the differences in the signal-to-noise ratio in both 'NG' and 'Confirm' were the largest among the others. After the signal-to-noise ratios were normalized, the factor of blade angle obtained the value of 1 in both 'NG' and 'Confirm' responses consistently.

The effect generated from the other two factors can almost be neglected in the SP process. Not only could it be concluded that the blade angle is the dominant factor in the SP process, but the 70° blade angle was also found to be the most efficient and effective measure in improving the process performance, rather than 50° and 60° blade angles. Although the industrial practice of using a 60° blade angle is widely adopted worldwide, this study explored a new research area further to improve the process performance in the PCBA process.

6.6 Contributions and Managerial Implications

Three contributions were achieved according to the above study, which could enrich the research and development work in the field of smart manufacturing [3], PCBA process performance and electronics manufacturing can be summarized. First, this study presents a systematic method by emerging new technologies into SP and SPI production process to formulate the IMPIS model in the PCBA workflow. A PSA model which includes the most critical factors in the SP process is also stemmed as a result of the IMPIS model. The factors of pressure, speed, and blade angle should be adjusted and fine-tuned in the SP process to obtain the improved yield performance.

Second, an exploration of the effect of using a 70° blade angle was undertaken, and such a 70° blade angle may become the new industrial practice in the PCBA process. Third, the proposed system (IMPIS) provides a flexible and robust approach to adjust the machines' parameter settings to obtain the optimal yield performance and maintain the quality of PCBs.

The measures established in this study are also aligned with the objectives of smart manufacturing and strengthen the competitiveness of PCB manufacturers. Consequently, the revolution of PCB manufacturing businesses can be further achieved through intelligence and digitalization.

6.7 Conclusions

Following an in-depth investigation on the background of the PCBA process and the motivation in the EMS, the critical factors and machine parameters in the PCBA process were identified, and yield performances were being improved. The proposed system, namely the IMPIS, which integrates of k-means clustering, IoT with seamless of data exchange between SP and SPI process, CCD, artificial intelligence, and Big Data analytics as a whole to fill the research as mentioned earlier gaps, was presented. To validate the proposed method, a case study in the SP process was performed. In this case study, the critical factors of pressure, speed, and blade angle were identified, and the optimal parameter settings were also determined. Along with this study, the PSA model was then established for the improvement of the SP process, and the 70° blade angle was found to be the most dominant factor in the process. Accordingly, the proposed system is not only beneficial to PCBA operations but is also valuable to the entire EMS industry to formulate strategies in smart manufacturing. Future work should focus on the exploration of more blade angles in the SP process and should replicate the proposed system about the other PCBA components.

References

1. Kuo, C. F. J., Fang, T. Y., Lee, C. L., et al. Automated optical inspection system for surface mount device light emitting diodes. *J Intell Manuf* 2019; 30(2): 641–655.
2. Wei, C. C., Hsieh, P., & Chen, J. Automatic adjustment of thresholds via closed-loop feedback mechanism for solder paste inspection. *Int J Elect Commu Eng* 2019; 13(11): 726–730.
3. Kusiak, A. Smart manufacturing. *Int J of Prod Res* 2018; 56(1–2): 508–517.

Index